People, Peace and Power

People, Peace and Power

Conflict Transformation in Action

Diana Francis

Pluto  Press

LONDON • ANN ARBOR MI

First published 2002 by Pluto Press
345 Archway Road, London N6 5AA

www.plutobooks.com

British Library Cataloguing in Publication Data
A catalogue record for this book is available from the British Library

ISBN 9780745318363 hardback
ISBN 0745318363 hardback
ISBN 9780745318356 paperback
ISBN 0745318355 paperback

Library of Congress Cataloging in Publication Data
Francis, Diana.
   People, peace, and power : conflict transformation in action / Diana Francis.
      p. cm.
      ISBN 0–7453–1836–3 (hardback : alk. paper) — ISBN 0–7453–1835–5 (paperback : alk. paper)
   1. Peace movements. 2. Conflict management. 3. Human rights movements. 4. Human rights workers. 5. Pacifists. I. Title.
      JZ5574 .F73 2002
      303.6'6—dc21
                                                                    2001006336

10   9   8   7   6   5   4   3   2   1

Designed and produced for Pluto Press by
Chase Publishing Services, Fortescue, Sidmouth EX10 9QG
Typeset from disk by Stanford DTP Services, Towcester
Printed in the European Union by Antony Rowe, Eastbourne

Contents

List of Tables and Figures

TABLES

FIGURES

Acknowledgements

Thank you to the workshop participants, who have taught me so much, inspired me and made my work worthwhile. Thank you to the co-facilitators in whose company I have learned new skills and with whom I have shared both tears and laughter. Thank you to fellow members of the Committee for Conflict Transformation Support, with whom I am able to share experience and think. Thank you to my family and friends for their moral support while I laboured over this book. Thank you to Peter Ellis, my indexer, for all his care. And thank you to Anne Rogers, without whose unfailing encouragement and editorial support it would never have reached completion.

The book grew (belatedly) out of my doctoral thesis, written under the auspices of the Centre for Action Research in Professional Practice at Bath University. I remain grateful to my supervisors and fellow students there, who helped me to grow as a 'reflective practitioner'.

The excerpt from Adrienne Rich's poem is printed with the permission of the author and W.W. Norton & Company Inc.

Acronyms and Abbreviations

ANC   African National Congress
ASEAN   Association of South-East Asian Nations
CCTS   Committee for Conflict Transformation Support
CSCE   Conference for Security and Co-operation in Europe (OSCE after 1995)
CODEP   UK Network on Conflict, Development and Peace
FRY   Federal Republic of Yugoslavia
KFOR   Kosovo Force
KLA   Kosova Liberation Army
LDK   Democratic League of Kosova
LEAP   Leaveners Experimental Arts Project (founded by the Leaveners Quaker Youth Theatre)
MOST   an acronym spelling 'Bridge', a group at the Centre for Anti-war Action in Belgrade
NATO   North Atlantic Treaty Organisation
NGO   non-governmental organisation
OAU   Organisation of African Unity
OSCE   Organisation for Security and Co-operation in Europe (CSCE before 1995)
PLO   Palestine Liberation Organisation
UCK   'Kosova Liberation Army'
UK   United Kingdom of Great Britain and Northern Ireland
UN   United Nations
US   United States of America

Preface

The manuscript of this book was completed two days before the horrific attacks made on the World Trade Center and the Pentagon (11 September 2001). Apart from brief references in the first and last chapters, they are therefore not discussed. At first this seemed a serious omission but, upon reflection, I have come to realise that the difference between what I *would have* written and what I *have* written is negligible. In fact, my book's title signals its relevance; and its perspective, analysis and focus can be seen as explanatory of those dreadful events and their aftermath, and indicative of what needs to happen in response.

The words 'democracy' and 'civilisation' have been raised as a banner for war. It is time for the right location and exercise of power – and the nature of power itself – to be re-examined; time too for reconsideration of what constitutes 'civilisation', and of the nature and role of culture. The notion of justice has been brought into sharp relief, and never before has there been a greater need to demythologise violence. These issues are at the heart of what I have written. I hope therefore that as people across the world try to rediscover their power to make a difference, this book will be of use, and will uplift the work of those who use their lives to do just that.

Part I

Thinking about Conflict Transformation

1 Introduction

Conflict can be defined as the friction caused by difference, proximity and movement. Since people and their lives are, fortunately, not identical, isolated or static, conflict between them is inevitable: a sign of life. So often, however, when we think of conflict, we think of pain, misery and death, of the violence and war with which it is so often associated. I would argue that this association is not inevitable, but stems principally from the near-universal cultural orthodoxy that frames human relationships in competitive and dominatory, rather than co-operative, terms: eat or be eaten, beat or be beaten; an approach whose logical outcome is genocide, nuclear terror, star wars.

This underlying culture of domination (Eisler, 1990) permeates the world of economics as well as international relations. The two come together in the arms race, which continues unabated and is fuelled by the economic contest of the arms trade. The needs of ordinary people, so many of whom live harsh and squalid lives, and the health of the planet on which all depend, are marginal to the perceived interests of these vast systems, increasingly operating on a truly global scale.

When the Cold War ended around 1990, it might have been hoped that better days were coming, that disarmament could proceed apace, that adequate attention could at last be given to the needs of the poor and that a new, co-operative model of international relations might emerge. With hindsight, any such optimism can be seen as ignoring the realities of old geopolitical interests and new rivalries, of the dangers of power vacuums and of the social misery occasioned by the collapse of communism. In many unstable regions, ambitious demagogues grabbed the opportunity to promote crude nationalism as a substitute for lost economic and existential certainties. Narrow and exclusivist identities, based on either ethnicity or religion, became powerful weapons in the struggle for survival in the 'new world order'.

With the overthrow of old hegemonies came civil wars and dissolutions. Although the old Soviet Union was broken up without much bloodshed, the region has since been beset by wars, large and small, short and long, and the economic plight of most of its people is disastrous. The carnage in Chechnya has horrified the world, though to no avail. Other violent conflicts in the region have passed almost unnoticed, yet continue to threaten lives and prosperity. In the former Yugoslavia, what Judith Large (1997) has termed 'the war next door' turned towns and villages into places of death and reminded Europeans that half a century after the Second World War peace and relative stability cannot be taken for granted.

Meanwhile, in Africa, the proxy wars in Angola, Mozambique, Ethiopia and other places, in which hundreds of thousands were killed, maimed and traumatised as the big powers struggled for control of the continent, were eclipsed in horror by the collapse of Somalia, the genocidal massacres in Burundi and Rwanda and the civil wars in Algeria, Liberia, Sierra Leone and Sudan. As I write, the war in the Democratic Republic of Congo threatens to engulf a large part of Africa. In Asia, the civil war in Sri Lanka seems endlessly intractable, while the winds of change and economic miseries brought about by the workings of the global market have destabilised the whole region, to the point of threatening the break-up of the old, despotic Indonesian empire.

With the old Soviet empire out of the way, and one superpower effectively unopposed, the underlying global contest is now more starkly revealed for what it is: the struggle for economic control and general world dominance. In the West, the 'need' for symbolic enemies is, meanwhile, supplied by the Arab world (often, but of course incorrectly, identified with 'Islam'), whose control of vital commodities makes its 'otherness' threatening and hateful. The ongoing bloody conflict in Israel/Palestine is arguably an old-style colonial war and the deadly terrorist attack on the United States in September 2001 was understood by many outside the West (and indeed within it) as a blow – however cruel and misguided – against the arrogance of neo-colonial tyranny.

Although issues of identity – tribe, nationality, ethnicity, religion – have been presented as the cause of so many recent wars, strategic interests and economic factors often play a fundamental role. In many cases, corrupt governments and rebel leaders use their power for personal gain. Politics and gangsterism are intertwined and the rule of law is weak or non-existent. In some cases this has led to the collapse of government, or its marginalisation, and makes the establishment of democratic systems exceedingly difficult. This is true in ever more parts of the world, including many former Soviet countries, Africa and Latin America, where violence is by no means the monopoly of states, but is widely used in economic activities, whether by individual gun-owners or private armies. At the same time, the influence – legal and illegal – of wealth over politics in many 'democratic' societies, is manifest, albeit dressed in the respectable clothes of the parliamentary lobbyist or party benefactor.

It is a sad irony that many decent people in the privileged West watch with distress the events which, like their own comfort, were, if not predetermined, prepared for and influenced by the conquest, demise and re-formation of past empires. The response of their governments remains a 'colonial' one, assuming the right and duty to control events by force, meting out punishment where their authority is questioned and perpetuating the view that violence and disregard for people's rights are acceptable from the powerful – in short, that might is right (Nederveen Pieterse, 1989).

Too small or 'atomised' in its manifestations and at the same time too large and general to be visible in all this is the most widespread tyranny of all, the oppression of women by men. This is not recognised as a 'conflict' or 'mass violence', because the direct violence takes the form of endless numbers of seemingly isolated acts. It is so ingrained and routine that it is invisible, or at least largely unrecognised by those who benefit from it or who have learnt to accommodate it. It is *the* relationship of domination, which, in its pervasiveness, overarches all others. In the midst of equal opportunities legislation and the rise of women in many spheres in the West, it is perhaps tempting to think that this has changed. Yet the reality is, for many, as bad as ever. In 'liberal societies', domestic violence and the economic and political marginalisation of women are still major problems, while in other parts of the world the situation is still worse, and the large majority of women are excluded from many forms of self-expression, participation and power, and suffer untold violence. The cultural, structural, psychological and physical violence to which women are subjected are part and parcel of the culture of domination, which women, alas, often help to perpetuate through the gendering and militarising of their children.

I am painting a bleak picture. Where is the hope? How can anything change if so much is wrong and the systems of domination are so strong and widespread? The paradox is that – in the midst of so much ugliness – kindness, decency and courage survive. People maintain or re-create their sense of dignity and meaning, they laugh, love each other, care about strangers and hope and even act for better things. Journalists in different parts of what was Yugoslavia risk their careers and physical safety by standing up for tolerance and decency. Human rights activists in Azerbaijan speak out against arbitrary arrest and detention. Community workers in Northern Ireland, voluntary and professional, devote their lives to overcoming the divisions that have bedevilled their society. In different parts of Africa, women campaign for property rights, and for a voice in discussions about war and peace. In Israel/Palestine, people of all ages work for democracy within their own societies and for understanding between them. In the South Pacific there is a region-wide campaign against nuclear pollution. In Colombia, men and women defy competing mafias by declaring their village a peace zone. In England, where I live, the old anti-militarist movement, though small at present, continues in the Trident Ploughshares direct action movement, whose attempts to 'disarm' nuclear weapons systems have been upheld by juries in recent court cases as lawful.

New communications systems not only facilitate multinational business, but also offer new possibilities for organising global movements and coalitions. Perhaps the most notable is the growing movement against global capitalism; but the new international treaty against landmines, and the success of the anti-debt campaign, Jubilee 2000, were also the results of international networking.

Alongside this kind of campaigning and direct action a new field has grown up, most often referred to as conflict resolution. It has been developed theoretically (starting in the 1950s and 1960s) by academics from the fields of international relations and organisational management, who have also involved themselves in practice. In social movements and at the interpersonal level it has emerged in various forms of 'alternative dispute resolution' and 'neighbourhood mediation' – particularly in the English-speaking world. In contrast with the moral partisanship of disarmament and justice movements and their confrontational style, this approach is based on the notion of impartiality and quiet diplomacy, and the idea of resolving conflict through a process of dialogue and problem-solving designed to address the needs of all parties. Although this is a modern field, it has traditional counterparts in many cultures. Its focus is on ending conflict and restoring relationships.

The implied emphasis on avoiding or ending conflict associated with 'conflict resolution' (not helped by such terms as 'conflict prevention') has been a matter of concern for critics. Mindful of the need to address underlying structural and cultural violence and of the inevitability of conflict in the process of change, they coined the phrase 'conflict transformation' – which is my chosen term. It is used to denote a whole collection of processes and their results: processes aimed at making relationships more just, meeting the needs of all, allowing for the full participation and dignity of all; processes through which conflict may be addressed without violence and either resolved (conflict resolution in the more specific sense) or at least managed (that is, kept within manageable boundaries and with its destructive effects minimised); processes through which hurt and hatred may be mitigated and even overcome, and coexistence made possible; processes for developing a 'constructive conflict culture' (Francis and Ropers, 1997), so that new and ongoing conflicts do not become destructive, but are able to contribute to the well-being of a society.

One of the influences which discourages most people, most of the time, from taking any form of social or political action is the culture of domination which, while it glorifies violence, incorporates the assumption that it is the task of some to rule and of others to be ruled. The culture that produces militarism and military machines, so often used 'in defence of democracy', also produces passive populations who do not participate in their own rule, even where the legal space exists for them to do so. Militarism and democracy are mutually contradictory. 'Conflict transformation', therefore, in the widest sense, will entail not only the shift of specific conflicts from the arena of violence into that of democratic politics, based on the rule of law, but also the transformation of cultural assumptions about the exercise of power: the substitution of *power with* for *power over*, and the assumption of responsibility by 'ordinary people', individually and collectively, for the things that affect their own lives and those of others.

The older field of Gandhian nonviolence has much to contribute to the theory and practice of conflict transformation. One of Gandhi's most important contributions was to revolutionise the theory and practice of power. His *satyagraha*, usually translated as 'truth force', is inseparable from the twin notion of *ahimsa* or 'non-harm'. It represents power as a moral energy, the ability to transform minds and relationships, rather than the capacity to control or dominate through the use or threat of violence. While those who work in social movements are subject to forces beyond their immediate reach or control, and influences of which they may not be aware, they also have many sources of power for transforming the world around them.

To work for conflict transformation at any level, therefore, involves ensuring that those who have been the subjects of structures of domination discover and develop the power to participate in what affects them. It means enacting democracy at all levels of public life: international, national and local, working in ways that increase participation and help people in all sectors of society to find a voice. It means supporting 'people power'.

The transformation of culture and of social, political and economic structures is an ambitious and long-term project. In the meantime, crises that occur within and because of current attitudes, relationships and systems, demand an immediate response. The domestic and international systems for reducing conflicts at an early stage are often so weak that intervention is only seriously considered when a full-blown 'crisis' has developed. It is argued in many circumstances that military intervention is the only effective means to quell large-scale violence. I believe that a careful review of recent military interventions, even in their relatively short-term outcomes, would largely not support this position. Decisions to intervene militarily are often based on self-interest and on a need to be seen as capable of exercising power effectively, rather than on a more objective assessment of the short and long-term needs of a situation. In any case, I would suggest that pacifists and non-pacifists alike, who are committed to the reduction of human cruelty and suffering, can agree that it is desirable to reduce to the minimum the military and coercive component of any response, and to help maximise the role and effectiveness of those local initiatives which contribute to forms of human security that are self-supportive and sustainable. To the extent that this is done, not only will long-term peace have a better chance, but the wider aim of getting beyond the culture of war and creating new, democratic, nonviolent norms will be advanced.

At the top level of existing political hierarchies, the 'conflict transformation' approach in a crisis is constructive negotiation, with or without the help of a mediator or facilitator. Such negotiation and mediation can be done officially and publicly, or unofficially, behind the scenes. More often than not, presidents and prime ministers are seen to participate in such talks only when the ground has been thoroughly prepared by others. So in the case of the Middle East peace agreement of 1993 (since sadly disregarded and now near to death), the 'Oslo process' of extended, informal and 'off-the-record' exploratory meetings played a key role, and 'problem-solving workshops' held under different auspices over many years with influential people of different categories contributed to the thinking on which the Oslo process was then able to draw. In Northern Ireland, behind the scenes negotiations had clearly been going on for many years before 'the peace process' became official in the early 1990s.

Secret negotiations at the top leadership level have the important advantages associated with confidentiality – the space to think experimentally, without having to justify every tentative idea to one's political constituency through the megaphone of the media, and the opportunity to build personal trust in the kind of relaxed atmosphere which is possible only in private. The dilemma is that secrecy has considerable disadvantages too. Shifts of perspective that take place behind closed doors through subtle processes in which others have not shared are not easily explained afterwards to the populations that will be affected by decisions based on them. The Middle East case offers a woefully clear example of the need for politicians to have the support of their constituencies when they reach agreements. With so much opposition to the Oslo Accord, it has proved impossible (or at least politically inexpedient) for the Israeli leadership to honour their side of the bargain, and extremely difficult for President Arafat and his supporters to hold the line on what was, for all Palestinians, a pretty poor deal, against the opposition and disruption of Hamas and other militants. And in Northern Ireland there is an ongoing question as to whether the constituencies of the different party leaders will continue to support the peace process if certain things (such as arms decommissioning or proposed changes in policing) do or do not happen.

The transformation of attitudes to conflict and human security requires changes in the assumptions, structures and practices which express and inform governmental and intergovernmental policy and organisation, and the handling of large-scale political crises will involve governments (as well as rebel leaders). Nonetheless, it is clear that, even in the relatively undemocratic societies we are accustomed to, people who are politically active at 'lower' levels can play a vital role, one which, from a democratic perspective, should be strengthened. The focus of this book, therefore, is on the role of non-military, non-governmental actors who want to work for justice nonviolently; to act as peace constituencies in situations of war or open political conflict; to become peace-builders in societies where violence, hatred, mistrust and antagonism have become the norm, where inter-communal relationships and structures are fractured or exclude certain groups, or where the rule of law and democratic processes have broken down. Their power may be limited, their lives and actions prey to circumstances beyond their control and to the actions of 'leaders' whose agendas they do not share, but they can also have an impact, at every level.

The idea that popular power could be exercised nonviolently was inspired by the life and example of Mohandas Gandhi and his followers, and by the Civil Rights campaign in the US in the 1960s, led by Martin Luther King. It was also developed in the thinking of sociologists and educators like Illich and Freire. The term *poder popular*, which had been used in the Guinean, Angolan and Mozambican violent liberation struggles of the 1960s and 1970s, was translated into 'people power' in the nonviolent uprising in the Philippines when, in 1986, unarmed crowds thronged the streets, blocking the tanks of President Marcos and precipitating the overthrow of a long-time tyrant. It was followed by popular, unarmed uprisings in Bangladesh and Nepal; then in central and eastern Europe, in the late 1980s and early 1990s, which changed the face of the world. The recent overthrow of Slobodan Milosevic in Serbia has shown that people power is alive and well.

The failure of a ten-year campaign in Kosovo/a (written thus to acknowledge both Serb- and Albanian-language spellings and the contested status of the territory) to overcome discrimination and oppression (see Clark, 2000) is a reminder that the large-scale success of popular action is dependent on many factors (some at least within the control of the activists) and cannot be taken for granted. But it is important to question the perspective that sees history in terms of governments, boundaries and treaties alone, or of shifting alliances and the rise and fall of empires, or of the structures and mechanisms of economic power, or even in terms of war and peace. Reality consists also in the lives that are lived by individuals and communities, in particular acts of cruelty or kindness, in the local affirmation or denial of solidarity, in one person's protection or betrayal of another.

The range of opportunities for constructive action is reduced during a violent crisis. It is greater when direct violence is still only a possibility rather than a reality, or once it has subsided. On the eve of war, when the dynamics of intimidation and killing are already under way, the efforts of those who oppose violence may fail to prevent it – as did, for instance, those of Women in Black and the Centre for Anti-war Action in Belgrade, the Centre for Peace, Nonviolence and Human Rights in Osijek and the Anti-war Campaign Croatia in Zagreb, in 1992. Nonetheless, those organisations did, and have continued to do, work that made and makes a difference, contributing to human rights protection, even during the war, providing training and education for constructive approaches to conflict, helping to build bridges between polarised groups and to restore working relationships. They have created small circles of influence and action which will go on widening, changing the ways in which people think and act.

PEOPLE POWER IN ACTION

Conflicts involve actors of several kinds: the immediate protagonists in the conflict; those who have influence upon them, including constituencies for different postures, processes and outcomes; bridge-builders and mediators (not to mention arms dealers, black-marketeers and extortionists, *agents provocateurs* and demagogues who manipulate conflict for their own purposes). It is usual for those concerned with the resolution of conflict to focus on the role of those whose aim is to build bridges and achieve a negotiated settlement, whether through community relations work or mediation or building a peace constituency. I would argue that it is not only those on the fringes of the conflict or playing an intermediary role who can contribute to a constructive outcome. The role of the protagonists is primary and can be constructive. The operation of 'people power' in addressing conflict constructively can, then, take many forms.

Nonviolent Action to Confront Injustice

Even when people appear to face overwhelming odds, they can act to challenge the power which oppresses them. The twentieth century provided many striking examples of this, some of which have been referred to above:

- the campaign led by Mohandas Gandhi to achieve India's independence from Britain
- the nonviolent action by Philippinos to overthrow the tyranny of President Marcos (captured on our television screens as nuns sat in front of tanks and offered the soldiers flowers)

- the school and rent boycotts in South African townships, which finally began to shake the pillars of apartheid
- the disintegration of the communist system in Europe (and Eurasia), when unarmed people took to the streets and insisted that things had to change.

The primary actors in all these cases were those who were the victims of oppression. However, support from outside played a part: mill workers in the north of England supported Gandhi and his followers; those in South Africa who were working for the removal of apartheid were supported by solidarity movements, boycotts and sanctions; and nonviolent activists in the Philippines had received nonviolence training from people with experience in other parts of the world.

Cross-party/Bridge-building Work by 'Insiders'

People from conflicting groups can take action together to represent common interest or to create channels for constructive communication and understanding. Some of the following examples are well known, others less so:

- 'community relations' work and behind the scenes mediation in Northern Ireland which over many years prepared the ground for peace
- the continuing dialogue in Israel/Palestine between those who understand that peace will come only when justice is assured for Palestinians, as well as security for Israelis
- women's action in north-east Kenya which brought the traditional leaders of feuding tribes together in a dialogue and led to an end to the fighting
- groups in Hungary that organised play schemes to bring children and parents from different ethnic communities together
- Radio Kontact – an inter-ethnic radio station in Kosovo/a, which broadcasts in different languages, letting different voices be heard; with similar media initiatives in Afghanistan, Burundi and elsewhere.

Intermediary Work by Outsiders

People working for external organisations may have a particular role to play in facilitating the re-establishment of constructive communication when a situation is very tense. Sometimes this work is done with groups of people, often in workshops. At other times it is inter-mediary work between individuals, as in the first example below:

- unofficial political mediation by Quakers in the Biafra–Nigeria civil war, in the anti-colonial war in Zimbabwe, in Sri Lanka and elsewhere
- the organisation and facilitation of workshops for dialogue and 'problem-solving' in the Middle East which prepared future government members and others on the way to the peace agreement currently in disarray; also the 'Oslo process' which led directly to the agreement
- the organisation and facilitation of dialogue workshops for young leaders of all ethnicities about the political future of the Balkans
- the organisation and facilitation of training workshops for women from Georgia and the seceded territory of Abkhazia, and problem-solving workshops for their political leaders.

Protecting Human Rights – Acting to Control Violence

Large-scale protection needs to be undertaken by regional intergov-ernmental bodies (such as the Organisation for Security and Co-operation in Europe (OSCE), the Organisation of African Unity (OAU), the Association of South-East Asian Nations (ASEAN)) and the United Nations (UN)). However, there are also examples of small-scale non-governmental intervention to protect human rights:

- peace and human rights activists in Osijek, Croatia, risking their own safety by going to stay with people threatened with violence and eviction from their homes
- multi-ethnic peace teams, with an international component, organised by the same group in Osijek, helping to manage the conflicts associated with the return of refugees
- individuals in Uganda, during the civil war in the days of Idi Amin, taking into their homes neighbours who were threatened with murder and refusing to give them up
- people in Colombia declaring their territory a peace zone and refusing to co-operate with any of the armed factions
- the work of Peace Brigades International, with team members acting as 'nonviolent bodyguards', for instance, in Guatemala, protecting those campaigning against political 'disappear-ances', in Honduras and Southern Mexico, accompanying

returning refugees, currently in Colombia, helping to keep a 'space' for human rights activists to operate (though recently under threat themselves).

Building a Peace Constituency

When war or widespread violence is under way, citizens may still act to influence their political representatives and governments to try and bring hostilities to an end. This is a specific form of nonviolent action that involves all the skills of movement-building and designing action to capture public attention. Although they may not always succeed, in the short term at least, they have the potential to do so if they can gain enough support. In the meantime they can help counter more hawkish tendencies. Some examples of such work are:

- the campaign of Peace Now and other organisations in Israel
- the movement to end Russia's war in Chechnya in 1995, led by the mothers of conscripts to the Russian army
- women's and churches' organisations in Sierra Leone campaigning for democratic elections in 1995/96 as a stepping stone towards peace
- the anti-war movements in the former Yugoslavia which maintained their activities from 1992 onwards, throughout the war and since
- the work of *Kacoke Madit* ('the big meeting'), an Acholi movement working for a peaceful solution to the conflict in Northern Uganda.

Education and Training

School lessons which develop tolerance and conflict-handling skills have been pioneered in many countries. Peace studies, under various names, constitute a growing academic field, and the centres which offer them also do important research work, analysing and collating the findings of experience. The role of education in supporting 'people power' is crucial. Its most usual vehicle is training workshops, which provide an opportunity for reflection and skills development, so helping people to recognise and build on their own capacities for peace. Some examples:

- the work of *MOST* ('Bridge') in Belgrade, formed in the 1990s, during the war – classwork in schools, training for teachers and the development of teaching materials; at the beginning, outside trainers worked with *MOST*'s members – now it is they who are the experts

- the nonviolence training which prepared people in the Philippines for the action which helped to oust President Marcos – done mostly by local organisations, although trainers from other countries made an initial input and continued for some years in a support role
- training for journalists in the run-up to elections in Nigeria in 1999
- training for women in India who work to overcome the violence associated with the caste system and the oppression and abuse of women.

THE NEEDS AND POSSIBILITIES OF DIFFERENT SITUATIONS

Where there is social stability but endemic or structural oppression or exclusion of one group or class by another, not only is there the current, if hidden, violence of injustice, but also the danger of future confrontation and overt violence. In such cases of latent or suppressed conflict there is the need for the creation of political awareness in disadvantaged groups and mobilisation by the oppressed themselves for peaceful change. Those outside the oppressed group may act as advocates for them and for their inclusion in political structures.

In societies characterised by social and political tensions which have not erupted into widespread violence but are in danger of doing so, interventions at the political level, aimed at encouraging political accommodation, constitute one potential form of conflict transformation. Interventions seeking to address the needs of disaffected groups, or to help them become more educated, organised and vocal would be another. (Here conflict transformation and development overlap.) Support for nonviolent action on their part, along with advocacy in relation to the dominating group, would be yet another. And bridge-building work between different communities can help to develop a constituency for peaceful coexistence and accommodation and against political violence.

Mediation may be helpful in a wide variety of contexts and at different levels, from the national political arena to local disputes. Certainly, different conflicting parties will need to be able to enter into negotiations if settlements are to be reached, and skills and principles for constructive negotiation will be needed. In all these cases, workshops can be used by those who live in the conflict area either as a vehicle or as preparation for what is being attempted on the ground.

Although it is very difficult for ordinary people to take any sort of constructive action when a conflict has resulted in widespread violence or war, many nonetheless have the courage to continue to work for the things they believe in, at whatever level they can. During the 1992–95 war in what had been Yugoslavia, citizens' groups all over the region were campaigning for an end to hostilities and for human rights, working with refugees and educating children in the principles and skills of nonviolent conflict-handling and inter-ethnic respect. Training workshops provided them with a great deal of support – moral and educational – and sometimes offered an opportunity for activists from different parts of the now divided region to meet and encourage each other, comparing notes and discussing their common goals and inevitable differences.

Dialogue workshops or meetings, whose main purpose is to discuss the issues which are being contested in the conflict, are hard to arrange when there is a high level of inter-communal violence. The polarisation which is manifested in and intensified by fighting is accompanied by intimidation against 'collaboration' of any kind, or any attempts to dilute enemy images. But even though at times when the conflict is particularly acute there may need to be a pause, as soon as there is a lull, there are likely to be people prepared to meet and talk – so long as care is given to context (including place and conceptual/political framework) and confidentiality.

At the higher political level the main focus for intervention during violent conflict will be on trying to establish dialogue which is focused on the search for a settlement. Usually this begins unofficially and secretly, often with the help of unofficial mediators (as in Northern Ireland), so that the political risks for the top leadership are postponed and reduced. Problem-solving workshops provide one kind of forum for unofficial, off-the-record dialogue which can feed into and ease the way towards official negotiations and help identify possible ways forward towards a political settlement. At the humanitarian level, training workshops can help those who plan and deliver aid to understand the ways in which they impact – both negatively and positively – on the conflict (Anderson, 1996; Miall *et al.*, 1999).

Once settlement of a conflict has been reached, the tasks of implementing the agreement and of social, economic and political recovery will extend into every level of society and affect every human life. Many of the issues will be the same as in a situation of tension which has not yet erupted into wide-scale violence; but there will be the added burden of hatred, fear and trauma which result from violence. The 'normalisation' of inter-communal relationships will be hard to achieve.

For a time at least, peace-keepers or monitors from outside may be needed – and training workshops may be part of their preparation. In any case, at the local level workshops will, as at other stages of conflict, provide a forum for dialogue and a means of training which can be supportive of all aspects of the work that needs to be done. They can help provide the ground and skills for addressing the ongoing conflicts that will continue to erupt at every level and in every sphere of social, economic and political life. They can also provide a safe place for the processing of hurt and anger, and a means of re-opening communication between embittered groups and facilitating the development of joint projects. The (re)building of democracy and civil society are likely to be important, and can be seen as integral to the post-settlement or post-violence stage of conflict transformation. Establishing mechanisms and a culture for the peaceful handling of conflict will be an essential part of this. It will be important for reconstruction and development work to be done in a 'conflict-conscious' way, fostering co-operation and equality and minimising the provocation of jealousies. Here again, training workshops have a role to play.

The different kinds of action that may be undertaken by 'ordinary people' at different stages or intensities of social/political conflict are summarised in Table 1.1 (p. 18). Workshops are included explicitly in many places, but they are also the vehicle for many of the named activities (such as bridge-building).

WORKSHOPS FOR TRAINING AND DIALOGUE

Training is the most vital means of supporting effective organisation and action, by multiplying the numbers of people with the awareness and skills required to act judiciously and have an impact. It can be facilitated by people in all three categories listed in Table 1.1 – partisan actors, cross-party workers and third parties – if they themselves have the necessary knowledge and skills. Workshops for dialogue and 'problem-solving' also have an educational aspect in that they help participants learn about each other and generate new ideas for effective action. Lederach (1995) argues that it makes sense

Table 1.1 The exercise of 'people power' at different stages of conflict

Stage of conflict	Partisan actors	Cross-party workers	External actors/'third parties'
Latent/suppressed conflict or oppression	• creation of political awareness and growing capacity for self-advocacy by disadvantaged groups • responsible media advocacy • acknowledgement and redress by dominant group • education and training	• development policy aimed at reducing ethnic stratification • promotion of multi-ethnic structures • bridge-building • education and training • careful media discussion	• monitoring of human rights and protection of minorities • encouraging creation of political awareness, capacity for conflict engagement and handling • acknowledgement of underprivileged parties • education and training • lobbying own governments to act constructively by supporting minority rights and development • facilitating dialogue
Confrontation/ chronic or sporadic violence	• working for nonviolent strategies • media stand against violence • mutual acknowledgement of legitimate concerns • direct negotiations • education and training	• bridge-building • promotion of multi-ethnic structures and loyalties • promotion of nonviolent strategies • media stand against violence and for dialogue • training for all the above	• monitoring and nonviolent protection • promotion and facilitation of constructive conflict resolution: round tables, pre-negotiations, workshops for dialogue mediation • establishment of institutions involving all parties • education and training • lobbying own governments to act constructively, by supporting dialogue and development

Wide-scale violence	• working for nonviolent strategies • seeking direct negotiations for an inclusive solution • media voice for ceasefire and negotiations • training for all the above	• bridge-building (where possible) • refusal to join in the violence • lobbying for nonviolent strategies • building support for peace • cross-party humanitarian work • media voice for ceasefire and negotiations • training for all the above	• monitoring and nonviolent protection • shuttle diplomacy and bridge-building • organisation and facilitation of pre-negotiations, problem-solving workshops • mediation • lobbying own governments to act constructively, by supporting dialogue
After wide-scale violence	• acknowledgement of responsibility; reparation • demilitarisation • participation in reconciliation process • education and training for peace • media support for peace culture • support for democratic process, government, law	• bridge-building • facilitation of reconciliation process • social rehabilitation and re-construction • promotion of a postwar peace culture (media) • promotion of multi-ethnic structures • education and training for peace	• monitoring and nonviolent protection • facilitation of reconciliation process • support in working through the traumas of the violence • education and training • lobbying own governments to act constructively, by supporting reconstruction and peace-building work and new democratic institutions

Source: originally published in Francis, 2001.

to concentrate these forms of intervention at the 'middle level' of society: those who have some influence, both 'upwards' and 'downwards', while being both free of the political constraints and pressures experienced by those at the highest levels of leadership and not so numerous as to render any meaningful impact unlikely. It is in practice most often at that middle social level that workshops are organised, particularly when they are resourced from outside the country in question. Those who participate are seen as potential 'multipliers', able to pass on what they have learned to those around them and increase the pool of people thinking in certain ways and willing and able to act. Often workshops are organised for relatively young, educated people who are likely to hold positions of influence in the future.

If the necessary skills exist within the group in question, training can be facilitated 'in-house' or by local trainers. However, it may be of benefit to have fresh perspectives and thinking brought from outside. The position of trainers and facilitators in relation to the conflict in question will also be important. To train partisan actors does not require impartiality (or partiality, for that matter), but it does call for a capacity to take some distance from the feelings and arguments at issue. As I have already suggested, where two conflicting parties are brought together for training, or where the primary purpose of workshops is dialogue between them, some emotional distance will certainly be vital for the facilitators, and they will need to be trusted as impartial or non-partisan third parties – either by virtue of belonging to none of the groups involved in the conflict or by being a balanced team of 'insider neutrals' (Hall, 1999).

Coming from outside has its own dangers, and the work of facilitation is full of pitfalls, requiring skill and sensitivity, as later chapters will suggest. The funding, organisation and facilitation of workshops of different kinds is in practice by far the most common form of supportive activity by outside non-governmental organisations (NGOs), and is often financed by governmental and intergovernmental bodies, as well as private foundations.

As the above section suggests, workshops are one of the most-used tools for conflict transformation. They are informal gatherings of people (usually not fewer than twelve and not more than 30) for a time of active thinking and talking together, with some inputs from those who facilitate or lead them, and using a variety of activities to stimulate thinking and encourage participation. They are designed for sharing knowledge, developing skills, learning new approaches and encountering 'the other'. They can support all those who wish to act constructively in conflict, whether from a partisan, semi-partisan or cross-party, or a non-partisan perspective (see Francis and Ropers, 1997).

Although workshops constitute such an important part of the practical work undertaken in this field, supporting the most legitimate power there is for conflict transformation, there is a relative dearth of literature about them and the challenges and dilemmas associated with their design, preparation, facilitation and evaluation. For that reason they will be given a central place in this book. As Table 1.1 suggests, outsiders go to 'conflict regions' to work in a variety of ways: as human rights observers; monitoring ceasefires; providing humanitarian assistance; and helping in peace-keeping and reconstruction once a conflict has been settled. Much development work is undertaken in areas suffering from violent conflict. Workshops are increasingly used to prepare people who undertake such work. My focus, however, is on workshops designed to support local actors (or activists) who live in situations of violent, or potentially violent, socio-political conflict.

All the different kinds of workshops alluded to in Table 1.1 can be seen as having one of two objectives (or both). One is to influence policy at the highest level, either by feeding in new thinking to the leadership via workshop participants or by helping to create a constituency for certain moves in policy, position-taking or decision-making. The other is to change the ways in which people think and act at the local level, affecting their own immediate environment and increasing capacities for social and political involvement. This second objective is often described as one of 'empowerment' or 'capacity-building'.

I shall continue my explanation under the two different headings: training workshops and dialogue workshops. I shall begin with training workshops and divide them into three broad categories.

Training Workshops

First, there are training workshops which bring together individual participants from different places: people whose social position, employment or personal interests and capacities give them the motivation and the possibility to be active in relation to conflict – usually in their home country but sometimes abroad. Such workshops can be local, national, regional or intercontinental. The wider their geographical scope, the further the balance of benefits shifts from those associated with a particular cultural or situational focus to those afforded by breadth of perspective, comparison and contrast. The hoped-for result of such workshops is that participants will gain new kinds of awareness and frameworks for understanding different events and contexts, and an increased capacity for constructive action in different situations (particularly their own). Individual participants may choose to keep in touch with each other after the workshop, or there may be some other kind of follow-up; but, as with most forms of education, the benefit of such a workshop is understood primarily in terms of individual learning – which in some cases can be immense and life-changing. The wider the geographical scope, however, the less likely it becomes that participants will continue to work together at the practical level, or that the workshop, through its participants, can have any appreciable direct impact on any one conflict situation. Such workshops can be seen as 'conflict intervention' only in the sense that they contribute to the pool of people with enhanced capacities for action in different areas of conflict, whether at home or abroad.

Secondly, there are training workshops organised for groups of people living within a particular place affected by conflict and seeking to increase their capacity to act, at whatever level. The benefits of such workshops lie in their specific, immediate focus, and the likelihood that they will result in increased effectiveness or new activities on the part of participants which, if well considered and well implemented, could have a positive impact on the conflict in question. The participant group will be likely to go on working together, or at least to maintain links, whether it is composed of members of one of the parties to the conflict or members of more than one party who nonetheless hold a common perspective and purpose. These workshops, like those in the first category, can be described as 'capacity-building workshops'. The capacities they are building may be for advocacy on behalf of a particular issue or group (for instance, work for human rights), nonviolent action in pursuit of some goal, negotiation, mediation, bridge-building between different communities, processes for reconciliation, and work to develop democratic structures and processes. In addition, they may help people to become trainers in all these areas, thus increasing the capability for capacity-building itself.

The third kind of training workshop brings together individuals from different sides in a given conflict who share a rather general will to counter the hostility and violence in which they are caught

up, but who bring with them different and often conflicting perspectives and experiences of the conflict. Such workshops have not
only a capacity-building but also a bridge-building purpose. They
have, in their goals and dynamics, much in common with so-called
problem-solving workshops. They can be difficult and explosive, but
can also result in breakthroughs at the personal and the conceptual
level. They make, in themselves, a small contribution to detente, to
deconstructing enmity, and they can give participants the tools and
the determination to go home and act for change, individually or in
concert, mobilising others to join them.

Dialogue Workshops

The difference between training workshops of this kind and
'dialogue workshops' is sometimes more a difference in emphasis or
presentation than in content, or a question of what is primary and
what is secondary. The methodology of workshops is always of
interest to participants, and in my experience they always claim to
have learnt from the new ways of thinking about and doing things
to which they have been introduced. In 'training workshops' these
methods constitute the main focus. In 'dialogue workshops' they are
vehicles for talking about and understanding the particular problem
or conflict in question, which is the primary focus of the workshop.
In situations where the latter kind of dialogue would be very difficult
psychologically (or politically impossible), it may be an excellent
alternative for participants to come together to discuss and learn
about some issues which are relevant to the conflict between them,
but are addressed in a more general way. It makes it possible to come
to the heart of the matter obliquely, and in good time, when the ice
has already been broken and participants have formed a community
of learning. They may spend the first few days of a workshop
applying skills and analytical tools to relatively 'safe' examples, and
turn to their own conflict only when they feel confident to do so. On
the other hand, in dialogue workshops the time and attention given
to the issues which have divided participants is more concentrated
and can more easily lead into further activities or action with a direct
relationship to the conflict. The very exercise of courage that
dialogue demands is important and, it could be argued, gives these
workshops greater immediate significance than training workshops,
whose impact lies in enhanced capacities for future action.

The kind of dialogue workshop about which most has been
written is the 'problem-solving' workshop. These are usually aimed

at people below the top decision-making level but with some political influence, ideally with direct access to those in power and sometimes even delegated to prepare the ground for them. Often facilitated by academics (see, for instance, Kelman and Cohen, 1976; Mitchell and Banks, 1996), they are designed to help the two sides see the conflict between them as a problem they need to solve together and the negotiation process as a joint effort to find ways of meeting the interests of each, rather than a battle over fixed positions. Many of the problem-solving workshops which have been written about were focused on the Palestinian/Israeli conflict, and several of those who participated have subsequently been appointed to government positions. The famous 'Oslo process', in which secret meetings held in Norway prepared the peace agreement which was signed by Yitzhak Rabin and Yasser Arafat, while it may not have been described as a 'problem-solving workshop', seems to have embodied the same kind of approach.

At the micro or local level, bridge-building workshops between different groups, whether they come in the guise of training or dialogue, can, like those high-level problem-solving workshops, have a discernible impact, though on a very different scale. For instance, a small project in East Slavonia, which runs such workshops in villages and small towns, has enabled former neighbours of Serb and Croat origin to face living together again, rediscovering each other's humanity and making a joint commitment to work for human rights and to organise income generation schemes. Several small local organisations have been formed in the wake of workshops, so they not only contribute to the overcoming of fear and resentment, but also play a part in the development of 'civil society' by facilitating active participation in social, cultural, political and economic life by ordinary citizens.

Although the knowledge, self-awareness and other skills needed for facilitating workshops can be acquired through participating in them and then simply starting to do it and learning 'on the hoof', their development can be accelerated through special training programmes for trainers and facilitators. This is arguably one of the most important forms of 'capacity-building'.

Johan Galtung (1990) describes a triangle of violence, in which direct, structural and cultural violence all enable and reinforce each other. Workshops for dialogue and training can, when their partici-pants have access to leadership or are able to translate new ideas into public pressure, have an impact at the highest political level where decisions are made for the ending of direct violence – 'peace-making' or 'negative peace'. They can also contribute to the deeper transformation of underlying conflicts, or 'peace-building', by creating the will and understanding needed for institutions and structural relationships to be changed. Much has been written about the dynamics of conflict (for example, Mennonite Conciliation Service, 1977; Glasl, 1997) and the mechanisms of polarisation and escalation which lead, in Glasl's eloquent phrase, 'together into the abyss'. Training workshops can enable their participants to become aware of these dynamics and to locate the points at which they can act to transform them. They can also help them to develop the skills and the confidence to do so. Dialogue workshops can address conflict dynamics directly, enabling their participants to transcend them. By challenging old assumptions and perceptions, both training and dialogue workshops can diminish the cultural violence which is used, in Galtung's words, 'to legitimise direct or structural violence' and make it 'look, even feel, right – or at least not wrong'. And they support people who wish to exercise their responsibility and power, with others, to play a part in shaping their own social and political reality.

TENDENCIES AND TERMINOLOGY

As I indicated earlier, the terminology within this field is confused and confusing. To summarise:

- The field itself is most widely known as 'conflict resolution'. 'Conflict resolution', as an approach and set of processes which led the field and gave it its name, is focused on mediated dialogue which seeks to address the fundamental needs of both or all parties to a conflict. It does not, as I suggested above and will discuss in Chapter 2, address major asymmetries of power and does not use the language of justice.
- The term 'conflict management' indicates an approach which assumes that the resolution of conflict is an unrealistic, utopian goal, and that the realistic approach is one that seeks to manage conflict in such a way that it does not become unnecessarily destructive. I believe that at times a conflict may need to be managed and at times the same or another conflict may need to be resolved.
- A 'conflict settlement', or political agreement, may be an important instrument at a given point in the management,

resolution and transformation of conflict. It may also be a step towards conflict resolution.

- 'Conflict transformation' embraces the different processes and approaches that are needed to address conflict constructively in different contexts and at different levels, in the short term and the long term, including engagement in conflict as well as its management and resolution. It is the term I shall use when I refer to my own perspective and to the field as I would like to see it, now and in the future. I shall, however, use the term 'conflict resolution' to refer to the field more generally or as it is often seen, as well as to indicate a particular set of processes within the overall scope of conflict transformation.

THE REST OF THIS BOOK

My purpose for the book is to bring together theory and practice. I shall write from the perspective of a practitioner who finds theory both fascinating and essential. This first part of the book is theoretical. Having set out my 'world view' and perspective on conflict, in Chapter 2 I will discuss the theoretical base for conflict transformation, arguing for a combination of conflict resolution and active nonviolence. In Chapter 3 I address the vexed question of the cultural transferability of that theory.

Part II will be focused on conflict transformation as prepared for and experienced in workshops. It will begin, in Chapter 4, with a general discussion of workshop content and methods. Chapters 5, 6 and 7 will provide detailed accounts of three very different training workshops, recorded from my own experience, and discussed in relation to the ideas set out in Part I. Chapter 8 will describe a series of dialogue workshops held in the changing context of events in the Balkans and centred on the relationship between Serbia and Kosovo/a. I shall conclude Part II with a chapter of reflections on the experiences of practice which have been described (Chapter 9).

Part III will be focused on the future of the conflict transformation field. In Chapter 10, I will discuss what I see as some elements of good practice and finally, in Chapter 11, I will consider a variety of cultural, conceptual and political challenges to the field of conflict transformation, arguing that if its vision is to have a future, it will entail radical change in the conduct of the world's major military powers.

2 Theory for Conflict Transformation

In this chapter I shall consider the theoretical elements necessary for a comprehensive approach to the transformation of conflict. I shall begin by addressing the criticisms of conflict resolution as a tool of neo-colonialism, relating it to the prevalent emphases of conflict resolution theory and suggesting that greater attention needs to be paid to questions of power and justice. I shall seek to demonstrate that the theory of active nonviolence provides a necessary complement to the insights of conflict resolution, and that the combined strengths of both can provide the breadth of understanding and range of resources needed for a comprehensive approach to conflict transformation.

Many writers in what is commonly known as the conflict resolution field have acknowledged the importance of embracing conflict as both an agent and outcome of change, recognising the demands of justice and the realities of power. Within the literature generally they are acknowledged. As early as 1971, Adam Curle conceptualised and depicted the way oppressive relationships are changed through increased awareness leading to confrontation. Chris Mitchell (1991), in his chapter on 'Recognising Conflict', draws on Johan Galtung's concept of 'structural violence' (Galtung, 1990). John Paul Lederach (1995) writes of the need for the empowerment of oppressed groups, and Louis Kriesberg (1998) develops the idea of 'constructive conflicts', in which escalation has to precede resolution. Yet the focus remains largely on the role of third parties – especially outsiders – as intermediaries, and the relatively little that has been written on training and on the role of inside actors, with few exceptions, confines itself almost entirely to mediatory or bridge-building roles. It is significant that the terms 'mediation' and 'conflict resolution' are often treated as synonymous. With the notable exception of Roger Fisher and William Ury (1981 onwards), mediation receives far more attention than negotiation, and preparation for nonviolent partisan action seems to be largely absent from the menu of conflict resolution theory and practice. The term 'conflict prevention' reveals the emphasis which is given to the avoidance of turbulence, as against activities for confronting and preventing violence of every kind and overcoming exploitation, discrimination, exclusion and oppression. It is therefore with some justice that conflict resolution is seen as an instrument of pacification – a new, more subtle weapon in the armoury of those who benefit from the status quo, rather than a means of achieving peace with justice.

What are the main elements of conflict resolution? What concepts and mechanisms does it offer for transforming destructive conflict into a process for constructive change? I shall confine my discussion to what Avruch (1998) has called the 'restricted' model of conflict resolution, in which the parties to the conflict decide its outcome, rather than having a 'solution' imposed from outside, and in which the role of 'third parties' is to help the parties find a mutually acceptable way forward.

ELEMENTS OF CONFLICT RESOLUTION

Conflict Analysis and Conflict Dynamics

Conflict resolution places great importance on the analysis of conflict: its history, recent causes and internal composition – the different parties, the nature of their involvement, their perspectives, positions and motivations, and the different relationships between them in terms of power, allegiance and interest – and the current conflict's evolution and dynamics.

Ronald Fisher (1993), for instance, outlines a sequence of escalatory stages, which begins with discussion (an optimistic viewpoint? In my experience, conflicts often begin with an absence of discussion when it could be most productive) and goes on to polarisation, segregation and destruction. Chris Mitchell (1981) elaborates the idea of conflict cycles, in which behaviour is the trigger for hardening attitudes and their effects, which in turn give rise to more substantive issues and more negative behaviours and a further escalation in hostility and in the level of coercion employed by the parties. Friedrich Glasl (1997) delineates nine steps of escalation, depicted as a downward staircase, whose final step is described as 'together into the abyss'. Glasl's is a brilliant account of the irrational and counterproductive ways in which conflicts evolve. It should convince anyone that the management and resolution of conflict require attention not only to the issues at stake, but also to psychological needs and processes.

The aim of any conflict intervention will be to help in countering the dynamics of escalation, contributing to a reduction in the level of violence or hostility and making space for dialogue. Although different theorists and practitioners may privilege one or the other, most would agree that handling conflict constructively involves attention both to the material and psychological aspects of conflict and to the conflict and the metaconflict. There is some argument about the importance of the relationships between the negotiators representing different parties and the desirability of rapprochement between them (see, for instance, Camplisson and Hall, 1996, who argue that if there is too much rapprochement between leaders, they may lose touch with the perspectives and demands of their con- stituencies). However, it is generally agreed that they need to recognise that they depend on each other for a solution and that they will need to co-operate in order to reach a settlement which meets their various interests. Mutual respect and an understanding of each other's position and pressures will make this co-operation more feasible.

Needs Theory

Of great influence in the development of conflict resolution theory has been the 'needs theory' of John Burton (1990 and in many other publications), which argues that unmet needs are the most frequent and serious cause of conflict, and that there will be no resolution without those needs being met. If the parties themselves can recognise their own needs, they may also be able to see that they are unlikely to be met through armed conflict, or by an immovable adherence to fixed bargaining positions, but rather through a search for ways of meeting the needs of all those involved. This is often referred to as a 'win–win' approach. This expression can seem facile and insensitive, given the likelihood that in protracted and (especially) violent conflict, much will already have been lost. It can also seem to suggest that no sacrifices or trade-offs will have to be made, which is unrealistic. The language of needs (and perhaps even more of fears – see Cornelius and Faire, 1989) does, nonetheless, offer a way for people to connect with each other at a level of common human experience. In a world where demagogues have exploited and defiled the name of justice to generate episodes of violence in which any vestiges of justice, respect or care have been destroyed, the non-judgemental, compassionate language of needs has the power to cut through the rhetoric of blame and enmity, introducing in its place a recognition of mutual vulnerability and shared humanity.

Dialogue, Negotiation and the Role of Third Parties

Dialogue is at the heart of conflict resolution. The dialogue may be general, for the development of trust, understanding and a co-operative relationship, or focused on the search for agreement and described as negotiation. Whereas negotiation may take the form of hard bargaining, in which the protagonists use their relative power to gain advantage over each other, its purpose in conflict resolution is to discover or develop common ground and reach a mutually acceptable agreement, through a co-operative process rather than a contest. Conflict resolution in the Burtonian school seeks to redefine the conflict in question as a common problem and to engage both (or all) parties in a process of analysis and the search for a solution which satisfies the needs of each. The kind of 'principled negoti-ation' promoted by Fisher and Ury (1981), while described in different terms, in practice follows similar lines.

In much of the conflict resolution literature, negotiation is discussed from a third-party perspective. The popular image of conflict resolution is that of the tireless, resourceful mediator who somehow persuades the warring parties to see sense and averts or halts the worst of human tragedy. In the world of *realpolitik*, mediation usually means 'power mediation', in which someone acting for a powerful state or coalition of states 'persuades' the parties to accept an agreement. (Interestingly, Lord Owen, interviewed on BBC Radio 4 about Slobodan Milosevic's first appearance at The Hague Tribunal on 3 July 2001, referred to himself and others like him as 'international negotiators'.) The facilitative mediator of 'restricted' conflict resolution (Avruch, 1998) can have nothing to do with enforcement, though it has been argued that some form of coercive intervention may in certain circumstances pave the way for mediation in the restricted sense (Mitchell, 1993). Mitchell argues that a variety of different third-party actors will be needed to fulfil a variety of intermediary roles and functions, including convenor (initiator, advocate), enskiller (empowerer), guarantor, monitor and reconciler (Mitchell, 1993: 147). This broadening of focus is helpful, setting the specific role of mediators in a wider context and bringing the aspirational world of restricted conflict resolution together with the real world of now.

Just as 'power mediators' may 'bring parties to the table', facilitative mediators may play a role in preparing the ground for negotiations, through separate conversations with the different parties. This will often be done through 'off the record', 'non-official' processes, sometimes over a period of years (Curle, 1986; Williams and Williams, 1994). The informal nature of the processes and the absence of political profile or affiliation on the part of the mediators, their political powerlessness, are what fits them for their task, rendering them non-threatening and enabling them to be trusted by the different parties as having no axe to grind or to wield. They must remain strictly impartial in the process. The only advocacy role open to them is that of 'being an advocate for the process of conflict resolution' (Mitchell, 1993: 142). (Although the idea of turning to a mediator from outside the community in question may be counter-cultural, the 'traditional' model of using an inside authority figure could hardly work when different communities are in conflict with each other.)

Not only do mediators work with parties separately in order to prepare them for meeting and talking to each other; other processes may be used to prepare the ground for negotiations, whether official or unofficial. The kind of dialogue which prepares parties for formal negotiations is sometimes termed 'pre-negotiation'. One such process is the so-called 'problem-solving workshop' referred to in Chapter 1, in which middle-level leaders from the different sides are brought together in a process facilitated by outsiders to the conflict – typically practitioner academics. (Mitchell and Banks, 1996, give an excellent account of the approach, but there is a substantial literature on problem-solving workshops, both in general and in relation to specific conflicts.) Although Burton, the originator of these workshops, saw them as quite different in character from what he understood as mediation (see Avruch, 1998), if the kind of mediation in question is of the type described above, it is based on the same principles. I used the term 'problem-solving workshop' rather loosely in my opening chapter. For a more precise definition of the term as used by academic practitioners I can do no better than to quote Avruch's summary (1998: 85):

A third party (usually a number of people – a *panel* of topical and process experts) brings conflicting parties together in a neutral and unthreatening setting to help them analyse the deeply rooted or underlying causes of their conflict; to facilitate unhampered communication between them; and to encourage creative thinking about possible solutions – literally, to 'problem solve'.

The idea and form of such workshops is open to adaptation. Whereas Burton's practice was based on academic institutions, used academic facilitators and emphasised analysis and expertise, the assumption that analysis and the intellect are the most important key to the resolution of conflict is open to cultural and theoretical debate. It could be argued that the change in perceptions achieved through the workshop exchange is as important as analysis in allowing movement towards the resolution of differences. It is the task of the facilitators to devise and assist in a process which will help the different parties to reach a new understanding of each other and of the situation, and to explore together different options for the future. Just as both material and psychological needs will have to be met in any solution to the conflict, so they will have to be met in the dialogue process itself.

Although mediation is usually discussed in terms of persons and single processes, I believe that mediatory structures have an important role to play in the management and resolution of conflict, providing occasions, places and mechanisms through which dialogue is possible: recognised meeting places, inter-religious councils, local peace committees, inter-communal assemblies, cross-community organisations and regular meeting places. The development of such structures can assist local people in establishing patterns for the constructive management of conflict.

The Importance of Constituencies

Although decision-making will not come within the scope of the problem-solving workshop, since those with the power to make decisions would not normally have the freedom to participate in such informal group dialogue, the facilitators of such workshops can be seen as mediators of a 'pre-negotiation' process. Problem-solving workshops can be seen as part of the process not only of exploration of the possibilities for peaceful settlement, but of building a constituency for settlement. In complex political conflicts, the opposing views of different parties, and different motivations of a variety of internal groupings, make the achievement of a universally agreeable settlement extremely difficult (if not, in the last analysis, impossible). The Peace Agreement reached in Northern Ireland on Good Friday 1998 was achieved through a clear policy of involving all players through direct talks of one sort or another, and all political parties in the official process. Some voted against the agreement, and some formed new paramilitary organisations; but a majority went with the agreement. The difficulties which followed demonstrate the continuing pressures of trying to hold a peace constituency together through an extended process. The Oslo Middle East process and subsequent peace treaty included only the Palestine Liberation Organisation (PLO) and the Israeli Government. The dissatisfaction of so many Palestinians with the agreement made on their behalf, compounded by the non-implementation of the treaty, have made the 'peace' achieved extremely conflictual, violent and fragile. The whole area of constituency-building and inclusiveness in the process of achieving settlements deserves more study and practical attention.

Constituency-building for a particular outcome to conflict, or indeed for a particular process for addressing it, involves bridging gaps between different factions and different levels of society. Problem-solving workshops are typically designed for actors with some influence but who are situated somewhere below the level where ultimate decisions are taken. The notion of political and social levels of actors in conflict situations is frequently mentioned in the literature of conflict resolution. John Paul Lederach (who espouses the terminology and perspective of 'conflict transformation') represents these levels as a pyramid (1994), depicting a wide base of 'grass-roots' society, a middle layer of relatively influential people, and a very small top level of leadership. His argument is that those in the middle layer can have an impact on those above and below them. Ropers (1995), citing Lederach, makes his own divisions between what he terms the Realm of States on the one hand and the Realm of Societies on the other, and between, in each case, the micro level and the macro level – the micro level being that of particular activities and the macro level being that of structures and systems. He considers how action at the social level can support efforts at the political level, and gives examples. I believe that if further attention were given – by internal actors and outsiders alike – to the possibilities for mutual support, effectiveness in peace-making could be increased. For instance, in Northern Ireland, it might be assumed that bridge-building community relations work somehow prepares the ground for peace; but did it play any role in helping to make possible the recent ceasefire? Could grass-roots work of any kind have done anything to help the talks on their way and prevent the resumption of the bombing? Did it help bring about the resumption of talks? If so, to what degree and how? It would make sense to examine such issues and explore the possibility of increasing strategic leverage and support between peace-making efforts at different levels. A similar question could usefully be asked about 'horizontal' relationships – those between different categories of actor at the different levels: could they be better co-ordinated, and how do they relate to the views and power (potential and actual) of the socially and politically inactive majority?

Recovery after Violence

Stimulated by post-dictatorship processes in Latin America and the former communist world, and more recently in South Africa, there is increasing attention given to post-violence processes and social recovery. The overall process of social healing is often referred to as reconciliation (though Assefa (1993) uses that word to signify a whole approach to conflict and relationships). This process of recovery can begin once an agreement has been reached between the parties which meets (in relative if not absolute terms) the needs of all concerned, at both the practical and psychological level. Integral to the agreement will be provision for its scrupulous implementation and sound procedures for redress when it breaks down.

Ron Kraybill (1996), in his thoughtful discussion of what reconciliation requires, outlines the following elements:

- physical safety, for example, removing people from the site of conflict, inter-positioning personnel between warring parties, protective presence, monitoring
- social safety, and a context in which there is acceptance for the expression of the emotions occasioned by trauma and the opportunity to talk about what has happened in order to try and make sense of it
- the means of discovering, as far as possible, how or why particular events occurred, and the rediscovery of relative identities, with a degree of confidence which can allow for the admission of imperfections and diversity, together with an acknowledgement of interdependence and a return to the acceptance of risk implied by trust

- the possibility of restoring relationships, predicated on the success of restorative negotiations, that is, negotiations focused on needs rather than on blame and leading to what Kraybill calls 'restorative' (as against 'retributive') justice – apology and forgiveness may well have a role to play, but cannot be demanded, especially by well-meaning outsiders.

This list of ingredients for reconciliation is necessarily given sequentially, but in practice the processes outlined are ones which feed each other. Some easing of tension will need to precede negotiations (as discussed above) and the healing process will not be able to proceed to relative completion until all the issues of the conflict have been dealt with, and all underlying needs met. This may include measures of reparation (whose effect may be largely symbolic, but which satisfy to some degree the need for recognition of wrongs done). It will often mean a redistribution of power in different forms and the creation of new laws and institutions.

Recovery after conflict is of course a personal as well as social matter, and the work of psychological healing after trauma may be done through individual counselling – though this is arguably a very Western model, both unrealistic and inappropriate in many contexts. However, pioneering work has been done, for instance, by women's organisations in Rwanda and Sierra Leone, through group work at the community level. Apology and forgiveness are probably hardest of all, both personally and politically. Maybe easier are joint acts of mourning, confession or purification – though where one side can be seen as the clear victim this will hardly do.

When severe violence has been inflicted and endured, to process the resentments, hatreds and traumas of the past at the group or national level as well as among individuals is an immense and daunting task. Just as 'basic human needs' can never be absolutely satisfied, so demands for 'justice' can never be absolutely met. Nonetheless, without such processing, there is the danger that the bitterness that remains will erupt once more in the future, resulting in a new cycle of violence and counter-violence. It is, no doubt, too easy for outsiders to talk of forgiveness. Healing may take generations, but it can be helped rather than hindered. To think of reconciliation in the first place as the achievement of some kind of return to normality, and the capacity to do business again together, to move from violence to politics, is perhaps ambitious enough, given the extreme complexity and difficulty of the processes involved (Francis, 2000a).

Peace-building and Peace Maintenance; Prevention of New Rounds of Violence

Processes of reconciliation cannot, in practice, be separated from the wider tasks of peace-building. Whether in situations of latent conflict which has not yet erupted into violence, or in post-violence situations which remain volatile, work is needed to adjust and stabilise relationships and address the economic, political and infrastructural problems which could precipitate an outbreak or recurrence of violence. This can be equated with what Boutros Boutros-Ghali termed 'peace-building', the mesh of activities which will be needed in a process of recovery from tension and violence. The building and maintenance of peace after widespread violence will often require massive efforts to enable the return and resettlement of internally displaced people and refugees; the reconstruction of infrastructures and buildings, the rebuilding of economic life and of social networks and institutions, legal systems and politics. According to Assefa (1993), what is needed is a 'reconciliation politics', encouraging the building of consensus and looking for common ground, seeking not to exclude but to include; a re-ordering of relationships within society and a whole new concept of governance; the establishment of a social order which is characterised by what Curle (1971) describes as 'peaceful relationships'. And since relationships are never static, if peace is to last it will require ongoing attention to the maintenance of sound working relationships and structures at all levels and in all spheres of society.

In recent years, development agencies, including governmental ones, have been paying increasing attention to the impact of violent conflict – and responses to it – on the possibilities for sustainable development. Seeing that conflict often grows in the seedbed of deprivation and exclusion, they have also recognised the role of development programmes in helping remove the causes of physical violence. Milton J. Esman (1997) discusses not only the way in which 'violence can destabilise the environment in which the agencies operate', undermining their efforts, but also the damage that can be done to ethnic relations by the 'ill-considered provision' of foreign aid. These two sides of the conflict-development equation are clearly set out and powerfully illustrated in Mary Anderson's *Do No Harm* (1996). Anderson argues, in addition – again with telling illustrations – that the way in which aid and development programmes are implemented can have a positive impact on the handling of conflict, as well as not making things worse.

The development of good governance and political participation (often termed 'civil society development'), including pluralism and the public expression of differing points of view on public policy, is understood as necessary for the establishment of stable and prosperous societies. Along with programmes for education, economic development and economic and social inclusion, it can be seen as part of 'peace-building'. At the same time, activities to educate and empower disadvantaged groups disturb existing social, political and economic relationships and can lead to conflict, or bring conflict to the surface, so that developing structures, attitudes and skills for constructive conflict handling need to be part of the overall development programme.

(However, it is fair to ask whose view of development this describes and who provides the resources for its implementation. The constraints and pressures felt by poor countries often make the delivery of such changes impossible. Those who feel free to prescribe the changes do not necessarily (if ever) feel under any obligation to finance them, or to change the economic relationships that give them the power to dictate what should be done. Ironically, it is 'global apartheid' (Alexander, 1996) that gives the world powers the dominance that not only impoverishes the larger part of the world but enables them to set standards of justice which they themselves ignore. 'Structural adjustment', neo-liberalism and democratisation are imposed on many African and other countries by the International Monetary Fund, the World Bank and donor governments, regardless of the needs of ordinary people and of the likely consequences of their further impoverishment.)

THE NEED TO ADDRESS THE DEMANDS OF JUSTICE AND REALITIES OF POWER

Empowerment, participation and the idea of just relationships as an essential characteristic – or indeed a definition – of peace bring us back to the criticism of conflict resolution with which this chapter began: that it ignores the demands of justice and the realities of power. I would add that it also lays too much emphasis on the role of third parties and non-partisan action, particularly on the role of outsiders. Indeed, the emphasis on impartiality is so pronounced as to seem to imply that 'taking sides' is bad, leaving no room for moral judgements, or indeed for realistic assessments of the effects of major power asymmetries. Maybe this is because to take sides over the substance of a conflict is confused with taking sides against the well-being of one of the parties, but this need not be so. Whatever the reason, the cost is, I believe, an underemphasis on the potentially constructive roles of those directly involved in the conflict (the 'primary parties'), and the place for advocacy and solidarity roles for third parties (Francis and Ropers, 1997).

The fundamental weakness of 'restricted' conflict resolution seems to be that it assumes that parties can be persuaded to see their mutual dependency, regardless of their relative power. But as the history of South Africa would suggest, those whose power is over-whelming, and whose comfort is in no way disturbed by the misery they cause to others, have no awareness of their dependency on them and no interest in talking about change. In order for them to be willing to give up their dominating position, and to enter into a dialogue of interdependent equals, there needs to be a shift in the 'balance of power'. Power takes many forms. It is not, as Boulding (1978) points out, only coercive power that operates in human rela-tionships. Changes in moral climate have their impact, as well as economic and political pressure. Still, those changes are brought about by human activity. They do not simply occur. Human activity is never devoid of power relationships and dynamics. And for just peace to be fully realisable in any part of the world, there will need to be a global realignment of power.

Within a process of dialogue, it is possible that a facilitative mediator can create the space for weaker voices to be heard and clarify whether any settlement is genuinely agreed by both or all sides rather than imposed by one of them; but it is improbable that a party with overwhelming power would enter into such a process, or make agreements on such a basis. Similarly, in 'problem-solving workshops' (which bring together influential members of opposing parties who do not have decision powers but may contribute to decisions), the facilitators aim to create an atmosphere and maintain a process in which any external power imbalances are excluded from the workshop process and participants are treated with 'parity of esteem'. This may be possible with the kind of middle-level leadership usually involved, who have the freedom that comes with the absence of ultimate responsibility. If, however, this parity within the process is too many miles distant from the external realities of power (popular attitudes, the dispositions of power holders and their power in relation to each other), the chances of any immediate applicability of the understandings reached in the workshop will be slim. If one party will not be seriously affected by the continuation of the conflict or the potential actions of their adversaries, the power holders in that party will have little incentive to listen to proposals from lower down which take the needs of weaker opponents seriously. In such circumstances, real negotiations, freely entered into and aimed at genuine and willing agreement, cannot realistically be expected.

(It is perhaps worth commenting here that the 'balance' needed for a constructive resolution process is complex, involving not only the relationship between the parties, but the extent of their disagreement, or the size of the demands that one party is making of another. To take a simple example, if factory workers are asking for an additional five minutes for their tea break, they will not need such strong union support as if they were asking for their working week to be halved. A more complex example can be taken from the years of Albanian resistance in Kosovo/a. When in 1996 Adem Demaci (later the main political representative of the UCK or 'Kosova Liberation Army') was calling for heightened nonviolent confrontation between the Albanian population of Kosovo/a and the Milosevic regime, he proposed at the same time a softening of the Albanian position, demanding, instead of independent statehood, exploration of models for equal status with Serbia and Montenegro within the Federal Republic of Yugoslavia (FRY). Thus he was aiming both to increase his side's leverage, and to decrease the gap between the positions of the different parties to the conflict and therefore the amount of movement necessary (see Clark, 2000).)

As so often happens in practice, 'talks' may be brought about and forced to some kind of conclusion by 'power mediation' – the intervention of a powerful third party, to 'bring the parties to the table', acting as a lever on behalf of the weaker side or in favour of a certain outcome. However, since that outcome will be an agreement judged by the third party to be either equitable or in line with its own interests, the 'empowerment' offered by power mediation is the empowerment of the mediator. This does not constitute conflict resolution in the ideal or 'restricted' sense. The parties themselves have little real choice and the resulting 'agreement', while it may be preferable to a continuation of war, will be, in all likelihood, a poor basis for lasting peace or the eventual healing of relationships. While power mediation can impose settlements, it cannot 'resolve' conflicts. It is part of the pattern of domination (Eisler, 1990), whose governing principle is that might is right, ignoring or overriding human dignity and need – if nothing else, the need of participation, the need to have choice about one's own destiny. There is a real danger that such imposed solutions will break down, or require such enforcement as to contradict any idea of what would normally be called peace. And at some time in the future new conflicts are likely to grow from the seeds of past suffering and hatred, and the sense of humiliation which comes from having had to dance to someone else's tune.

COMPLEMENTING CONFLICT RESOLUTION

If 'conflict transformation' is to be worthy of its name, it must address the limitations of 'conflict resolution' outlined above. It must have something radical to offer in conflicts where power asymmetry is not incidental but of the essence. It must do so, I would argue, not by recourse to the contradictory notion of power mediation or abandoning the field to military intervention, but by embracing the knowledge and experience, the values and the passion of active nonviolence, whose essence is to *engage* in conflict in order to resolve it. Nonviolence and conflict resolution are both concerned with action to overcome violence, transforming the dynamics of the conflict and establishing relationships of respect. But, whereas conflict resolution concentrates on minimising or ending what is often the secondary violence of armed, or otherwise hurtful, confrontation, the primary concern of nonviolence is to overcome the structural violence of injustice, by nonviolent means. And whereas in conflict resolution the focal actors are impartial third parties, in nonviolence they are parties to the conflict, acting in the first place on their own behalf, to rectify a situation of injustice, while respecting the humanity and needs of those whom they oppose.

My work as a trainer, facilitator and consultant in situations of conflict has brought me face to face with these issues at the level of praxis. The need to turn theory into practice led me to the realisation that a body of expertise already exists which is tailor-made for this purpose, ready to be adopted and combined with the theoretical framework, the psychological insights and the practical skills of conflict resolution, complementing it in such as way as to create a more comprehensive menu. As Miall *et al.* (1999) explain, some of the roots of conflict resolution are in the older nonviolence tradition, and many of those now working in the field once marched, as I did, under the banner of 'nonviolent struggle'. It is interesting to speculate as to why the two fields have remained to a large degree separate, in theory if not in practice. Maybe postmodern angst and the swing away from ideological certainties made for discomfort with the absolutist flavour of Gandhian nonviolence and its arguably extreme moral demands. Perhaps the analytical–psychological mix of conflict resolution is more in tune with our times than the leftist political flavour of nonviolent activism. I was drawn to the notion of conflict resolution in response to the rash of bloody conflicts which has disfigured the world in recent years, recognising that in justice's name great cruelties were being committed. I realised that I had been so preoccupied with the notion of justice that I had perhaps underestimated the dangers and costs of struggle and the need for accommodation. I saw that this new field of conflict resolution had a great deal to offer in situations already torn and wasted by antagonism and violence. I began to see that not every conflict could be understood as a struggle between oppressors and oppressed; that injustice and oppression can be used as slogans by demagogues motivated by personal ambition rather than respect. Nonetheless, injustice and oppression do exist; indeed, they characterise a substantial proportion of human relationships, and those relationships need to be transformed.

Whatever the cause for their separation, therefore, I believe these two fields of nonviolence and conflict resolution need to discover that they are not just nodding acquaintances but blood relatives, in fact the twin halves of conflict transformation. Having discussed, albeit briefly, what I take to be the central elements of conflict resolution, I will attempt to summarise the philosophy and methodology of nonviolence, drawing on the thinking not only of Mohandas Gandhi, but of others whose thinking has contributed to the nonviolence movement as it has evolved.

NONVIOLENCE AS A PHILOSOPHY

Gandhi developed his philosophy, system of action and actual campaigns in response to the injustice of colonialism. Through his resistance to the imposition of apartheid in South Africa, in the first two decades of the 1900s, and his subsequent leadership of the struggle to free India of British rule, he demonstrated and developed a cogent, coherent approach for responding to violence and injustice and for acting to establish what he considered to be wholesome patterns of life style and relationships. In an undated book entitled

Non-Violence: Weapon of the Brave, Gandhi wrote (p. 15), 'I believe myself to be an idealist and also a practical man.' His idealism was based on the twin concepts of *satyagraha* or 'truth-force', and *ahimsa* or 'non-harm', his approach was rooted in Indian Hindu tradition and reached into every aspect of social and personal life. (I often use the phrase 'active nonviolence' to reflect this combination of non-harm with positive energy and comprehensive activity.)

Gandhi also frequently acknowledged his debt to the teachings of Jesus. In the secularised West, it is easy to forget that most of the world's people are still religious, and that to ignore the spiritual dimension of their lives is to disregard and disrespect an essential part of their identity and motivation. Gandhi's close colleague Abdul Ghaffar Khan was a devout Muslim, and his followers called themselves the *Khudai Khidmatgar* – 'Servants of God'. The Civil Rights movement in the US was based in the Southern churches. The nonviolence movement in Latin America has been built on Catholic base communities, sometimes with a mix of indigenous beliefs. In Vietnam, opposition to the (partly) civil war was led by Buddhist monks. More recent demonstrations of nonviolent 'people power' have been inspired and strengthened by the religious convictions and inner preparations of their leaders and groups – for example, in the Philippines, in South Africa, in Eastern Europe and in the Baltic States where, although the masses of demonstrators were (with the exception of Poland) for the most part not religious, the churches often played a catalytic role and provided an umbrella for meeting and joint action.

Although many Western adherents of nonviolence are not religious, in the words of Jean and Hildedegard Goss-Mayr, 'The nonviolent struggle is essentially carried out at the level of conscience' (1990: 27). Nonviolence is not to be understood simply as a set of tools and tactics; not even a discrete field of human activity. In the thinking and practice of its founders, it was a way of life, inner as well as outer, through which the individual became capable of nonviolent action in particular circumstances, for particular goals. Both goals and methods were chosen according to strong and unequivocal values. Although Gandhi believed that absolute truth was the property of God alone (Bose, 1972), honesty and integrity were vital to his understanding of nonviolence. The pursuit of self-realisation, which is the basis for his ethic, presupposes the search for truth. Since all beings are ultimately one, violence against any living being is contrary to self-realisation, and 'permanent good can never be the outcome of untruth and violence'. There is 'the same inviolable connection between the means and the end as there is between the seed and the tree' (Gandhi 1980: 75). Processes and outcomes are inseparable; means and ends are one. The task of the self-realising individual is to act with others, nonviolently, for the reduction of violence (Naess, 1958). This understanding of self-realisation is close to the concept of respect, discussed further in Chapter 3, which I believe to be fundamental to conflict transformation and in which self-respect and respect for others are inseparable.

The world of active nonviolence, then, is a world of struggle, of protest, of action for change: change in power relations, in structures, in culture, in politics, and in the methods of struggle itself. From this perspective, conflict is seen largely in terms of justice, or the lack of it, and active nonviolence as the way to achieve it without at the same time denying it to others and negating the values on which the concept of justice is based. In nonviolence theory, to engage in conflict is not only in itself positive – it is a human obligation. According to Gandhi (1980: 81), 'No man could be actively non-violent and not rise against social injustice no matter where it occurred.'

Gandhi's nonviolence is based in egalitarianism. To him, socialism was 'a beautiful word', embodying a vision of a new reality, in which 'the prince and the peasant, the wealthy and the poor, the employer and the employee are all on the same level' (1980: 75). He saw the world as divided into haves and have-nots, powerful and disempowered, a division which needed to be replaced by equality and unity. For that to happen, the inequalities and injustices present in society had to be confronted, and the process was, in practice, highly conflictual. Gandhi and his followers defied the authority and the might of British imperial occupation. Similarly, the US Civil Rights movement constituted a deliberate engagement in conflict by black people and their supporters. When charged with creating conflict, Martin Luther King (Jnr) replied that the conflict already existed. What he and others were doing was to bring it into the open, so that it could be 'seen and dealt with' (King, 1963: 85; cf. Curle, 1971 on 'latent conflict').

ELEMENTS OF NONVIOLENCE

Conscientisation and Mobilisation

The first stage in a nonviolent campaign to address a system of injustice (Galtung's 'structural violence') is to awaken the self-respect of those subjected to it, and enable them to understand the nature of their condition. Such awakening was described later by the educator Paulo Freire as 'conscientisation', a process in which those involved:

> simultaneously reflecting on themselves and on the world, increase the scope of their perception ... [and] begin to direct their observations towards previously inconspicuous phenomena.

As a result of this new awareness:

> That which had existed objectively but had not been perceived in its deeper implications (if indeed it was perceived at all) begins to 'stand out', assuming the character of a problem and therefore of challenge. (Freire, 1972: 56–7).

This is the beginning of the process of 'empowerment', which makes possible the next step of organisation or mobilisation for action to change it.

The first step in mobilisation is group formation. This involves the clarification of values, the search for shared understanding through joint analysis of the existing situation, and the articulation of aspirations and goals. Through these processes trust and commitment are fostered and the capacity is developed for cohesion and joint action in the face of pressure and attack. Next comes the task of devising a strategy for the accomplishment of agreed goals and determining the methods that are to be used.

Dialogue and Direct Action to Confront Injustice

In active nonviolence, dialogue comes first and last. Gandhi's ideal was to convince opponents of the rightness of the cause which they were opposing and of essential shared interests around which co-operation was possible. Nonviolence meant fighting antagonisms, not antagonists. The methods of *satyagraha*, such as strikes or fasting, would be used only after the completion of other clearly articulated steps: the gathering and clarification of factual information; clarification of common interests; formulation of an interim goal which could meet the interests of both sides, followed, if necessary, by a further search for agreement through compromise on non-essentials (Naess, 1958). (It is perhaps worth noting here the similarity with Burton's needs theory and focus on interests, combined perhaps with a more explicit acknowledgement of the role of compromise.) The dignity of the opponent was, according to Gandhi, to be preserved at all times. The dialogue process was to be respectful and opponents were to be trusted and their truthfulness assumed. The nonviolent campaigners were to acknowledge the possibility that they were mistaken. They were also to be as transparent as possible about their reasons for doing what they did, and constantly available for talks. It was important for them to do all they could to understand and empathise with their opponent. Such understanding would lead them, for instance, to make concessions on non-essentials, in order to make it easier for the opponent to shift position, and to make their constructive intentions as clear as possible, in order to remove the adversary's suspicions.

But as Gandhi, King and others knew, those who benefit from oppressive or exploitative relationships do not readily enter into dialogue with a view to relinquishing their advantages. A change in power relations is likely to be needed before the process of dialogue will begin. Communication, therefore, while dialogue is refused by those in power, is directed towards potential friends and allies, with a view to building the support that is needed. As a movement grows, various forms of public action may be used to confront those in power, bring the injustice in question into the realm of public debate and build the pressure for change – moral, numerical, political, financial – increasing their leverage in relation to their demands. Gene Sharp, in *The Politics of Nonviolent Action* (1973), lists 197 forms of nonviolent action. They include petitions, marches, symbolic acts, blockades, occupations, strikes, boycotts, and civil disobedience. In practice the forms are infinite, limited only by the imagination of those devising them. The overall purpose of such actions is to bring about a situation in which the impact of the campaign and therefore the relative power of the campaigning group are such that it becomes worthwhile in the adversary's eyes to enter into dialogue with them.

Gandhi is clear that since the goal of nonviolence is self-actualisation and the overcoming of social division, the adversary must eventually be included in the 'permanent good' which is sought. Gandhi's thinking is again mirrored very closely by Freire's: 'This, then, is the great humanistic and historical task of the oppressed: to liberate themselves and their oppressors as well.'

In this way they would become 'restorers of the humanity of both' (Freire, 1972: 21). King, likewise, included white people in his dream for the future, proposing that he and his followers should act with enough love to 'transform an enemy into a friend' (King, 1969: 160).

Constructive Programme

Gandhi was concerned not only with the removal of oppressive laws and regimes but also with the development of alternative systems and lifestyles. He insisted that a parallel programme of social and economic development should accompany the struggle for liberty (Narayan, 1968). Ghaffar Khan, his Moslem colleague, required that his followers should do two hours of social work every day. To protest against something was not enough; it was vital to know what you wanted instead, and to work for it (Powers and Vogele, 1997). Thus the idea of empowerment promoted by nonviolence embraces not only action for political change but personal responsibility for positive action at the community level. Gandhi saw this 'constructive programme' as vital for the maintenance of a positive attitude amongst campaigners. This thinking seems particularly important if the idea of nonviolent struggle is applied to inter-ethnic conflict, suggesting an important way of helping to avoid the construction of ethnic identity in largely antagonistic terms, or in terms of victimhood. And in the aftermath of the collapse of the communist system, it is easy with hindsight to see the disastrous effects of the absence of visions, preparations and coalitions for the future. (Perhaps the most impressive example of a 'constructive programme' in recent times was the creation and maintenance of parallel institutions in Kosovo/a in the years preceding NATO action against Serbia – structures which were destroyed by the war and its aftermath (Clark, 2000).) A link can, I think, be made between the concept of 'constructive programme' and the idea of 'peace-building'.

DIFFERENT EMPHASIS, COMMON GOALS

In summary, it could be said that the overarching characteristic of nonviolence is its concern with the power and responsibility of 'ordinary people' to change the relationships which affect their lives and its assumption that their capacity to do so will depend on their own internal attitudes. Nonviolence is primarily a response to structural violence, embracing conflict as a means of change and developing a praxis of nonviolent action. It emphasises the importance of information and understanding of the other. Its goal is to create relationships which honour the humanity of all. Conflict resolution can be seen largely as a response to the violence and misery by which conflict is so often characterised, and constitutes a body of theory and methods whose main focus is on getting beyond confrontation. Psychological and political analysis, the mediatory role of third parties and respect for the needs of all parties can be seen as its hallmarks. At the same time there are increasingly loud and frequent calls from the conflict resolution field for more attention to be paid to 'conflict prevention' (in my terms, the prevention of widespread violence).

The essential values of the two fields, which can be summarised in terms of respect, are not only compatible but arguably almost identical, although one is more ideological/religious and the other more psychological/pragmatic in flavour. Nonviolence emphasises justice and conflict resolution concentrates on needs. Respect can be seen as bringing the two together, since respect for the needs of parties, and insistence on parity of esteem, is what can deliver justice. The focus on dialogue and on the systematic gathering of information is also shared.

The relationship between active nonviolence and conflict resolution can be seen sequentially. Nonviolence prepares disempowered and oppressed groups for conflict engagement. Conflict resolution comes into its own when the scene is ripe for dialogue. In addition, conflict resolution has contributed to our understanding of the way in which actors at the grass-roots and middle levels of society relate to what Ropers (1995) calls 'the realm of states' and to the terminology of the UN. It has generated much new thinking on what 'peace-building' requires after widespread violent conflict. It brings into the twenty-first century the age-old endeavour of seeking ways to overcome human destructiveness. Together, nonviolence and conflict resolution can provide a broad base for the continuing process of learning how to transform violent conflict into a constructive process of change. (Lest I should sound too utopian, I will add that the process of learning is in its infancy and the transformation of violent conflict is likely to present a challenge as far into the future as one can see.)

A SYNTHESISING DIAGRAM

For training purposes, in order to present what I felt to be a satisfactory synthesis of the two overall approaches, I developed a diagram. It had its beginnings when I was starting to work as a freelance facilitator and trainer, and wanted to explain the place of active nonviolence in working with conflict. It was based on a diagram I had seen in a book by Adam Curle (1971) in which he plotted the stages of conflict from the latent conflict of oppression, through the awakening of awareness and confrontation, to negotiation and sustainable peace. My own diagram was, I felt, easier to follow, being linear, and allowed for more elaboration. That rudimentary diagram became, in turn, the basis for the much fuller one I developed with a colleague, Guus Meijer, at a seminar in Nalchik, in the North Caucasus, in response to a request from participants that we should do something to address the question of power. They felt that our focus on dialogue did not match their realities as small peoples struggling for recognition in relation to the overarching power of Russia. We presented the new, more elaborate diagram under the title 'Power and conflict resolution: the wider picture'. It elicited an enthusiastic response from participants, who readily located themselves on it, and provided the basis for an exploration of the possibilities for increasing the power of oppressed groups, working on participants' own cases.

Since then I have used this diagram frequently (and it has been reproduced in various forms by others, for instance in International Alert's Resource Pack, 1996, in Miall *et al.*, 1999 and in Fisher *et al.*, 2000). It has provided a useful framework for my own thinking about conflict and for discussions with colleagues. It has also been useful in workshops to facilitate awareness and discussion of different stages of conflict – the challenges they present and the range of responses which can be made – and it has provided me with a framework for structuring and explaining the workshop agenda as a whole – what it is intended to include (and exclude) and why (though this is always open for discussion). The diagram is reproduced in Figure 2.1, with its accompanying text.

It is not meant to represent the one and only route into and through conflict, but is intended as one way of representing one possible route. It is a model for working through conflict – transforming it – from the hidden conflict (pre-conflict) of 'quiet' oppression or exclusion, up to and beyond open conflict or confrontation and its subsequent settlement, and through to the rebuilding and maintenance of a social infrastructure which can prevent the emergence of new forms of oppression and new outbreaks of destructive conflict. It does not, however, include all eventualities (for instance, what happens when one side wins outright and uses its victory to humiliate and marginalise, or even liquidate, the losers).

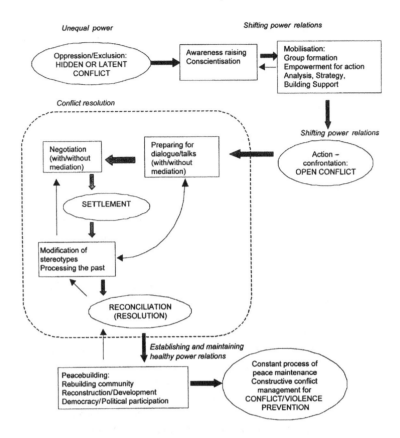

Figure 2.1 Stages and processes in conflict transformation

Stages and Processes in Conflict Transformation

The diagram in Figure 2.1 describes the stages and processes that will usually need to be passed through, if a situation of oppression, with an extreme imbalance of power, is to be transformed into one of genuine peace. (The words contained in the round or oval shapes describe conflict stages, while those contained in rectangles describe the actions or processes by which new stages are reached.) The 'stages' are not in themselves static. They have their own dynamics and in practice they may merge with one another. Neither are they likely to follow each other in a clear and orderly sequence. It will often be a case of 'two steps forward and one step back', or even vice versa, and frequently processes need to be repeated, built on or reinforced by other processes in order to bring about substantial progress.

In addition, large-scale conflicts are not simple or single affairs, but usually involve multiplicities of issues, parties and sub-parties. They will in all likelihood also involve conflicts and power struggles within as well as between parties, and the stages of these internal conflicts may well not coincide with the stage the overall conflict has reached. Nonetheless, this simplified diagram may provide a useful framework for thinking about the stages of conflict and shifting power relations.

The diagram begins with a situation in which the oppression (or exclusion) is so complete that the conflict is hidden or latent, the oppressed group remaining passive in the face of extreme injustice or structural violence (often maintained by physical violence or the threat of it). They may remain passive because of tradition or lack of awareness, or because the power imbalance is such that they have no chance of being taken seriously in any demands or requests they might make. In order for this to change, some individual or group will need to begin to reflect upon, understand and articulate what is happening and encourage others to do the same – a process described in the liberation language of Latin America as 'conscientisation'. This process will, if it generates sufficient determination, lead to the formation of groups committed to change. Their first task will be to continue the process of reflection and analysis, formulating a common purpose and strategy, then developing organisationally as they begin to take action to build support and so increase their relative power.

Some oppressed groups choose to use violence in their struggle, for others violence is not considered or is not seen as a practical option. Yet for others it is a matter of clear strategic choice or principle to act nonviolently. The term 'conflict transformation' implies the nonviolent option.

As their power and visibility increases, as their voice begins to be heard, these groups will increasingly be seen as a threat by those in power and a stage of open confrontation becomes inevitable – a stage which may well involve repressive measures, including physical violence, on the part of the oppressive power holders, even if the oppressed group have opted to act non-violently. During this stage of open conflict, the relationship in power between the opposing parties will change as a result of their ongoing confrontation and other developments inside the parties or in the wider environment. Even if the confrontation takes the form of armed conflict, eventually a road back to dialogue has to be found. Once the oppressed groups have increased their relative power or leverage sufficiently, they can expect to be taken seriously as partners in dialogue.

At this stage it is possible to begin the processes grouped together and described as 'conflict resolution', in which communications are somehow restored and settlements reached. This will not be a smooth process. Talks may break down, agreements may be broken, the conflict may flare up again. Non-partisan intervention can help – for instance in the form of mediation – both in preparing the parties for negotiation and in the negotiations themselves. And through the work of preparing the ground, and through face-to-face dialogue, some of the heat may be taken out of the situation, some more hope and trust generated, some of the prejudice dissipated, which in turn will facilitate the reaching of and adherence to agreements. Once these are in place, it may be possible to begin to deal with some of the remaining psychological damage which the conflict and its causes have occasioned and to develop more positive relationships between the previously conflicting groups.

These more positive relationships will be consolidated through a long-term process of peace-building, and will find expression in social, political and economic institutions. But societies never remain static and the final phase of 'peace' will need to be, in fact, a process (made up of a thousand processes) of maintaining awareness, of education, management of differences and adjustment and engagement at all levels, so that some new situation of oppression – or other major source of conflict – does not develop, and just and peaceful relationships are maintained.

Note: Extreme imbalances of power are not the only starting point for the route to open conflict. The stages and processes leading to it may begin elsewhere. But questions of power and justice need to be taken into account in any consideration of conflict and how to engage in or respond to it. On the one hand, the untimely 'resolution' of conflict may mean in practice the suppression of just aspirations – 'pacification' rather than 'peacemaking'. On the other hand, those wishing to enter into conflict in the name of a just cause need to do so with some understanding of the likely cost to all concerned, and of their current and future possibilities, in the light of the distribution of power.

No diagram can include all considerations, cover all eventualities, or represent all the ambiguities, nuances and complexities which characterise lived experience. In real life, as against diagrams – as I indicate above – stages and processes are not clear-cut and separate. They do not begin here or end there, they merge. They do not flow smoothly forward, but have their own unpredictable dynamics. They are unlikely to follow each other in a clear and orderly sequence. The growth of a movement for change is often a confused and unpredictable affair, prey to the whims, ambitions and manipulations of political entrepreneurs and demagogues, and swept aside or along by outside events beyond the control of those directly concerned.

Likewise, the impact and outcome of action taken will be affected by many extraneous, ungovernable forces, as well as by the choices of some or all of the members of a given group. The power struggle which ensues may send the oppressed group back to square one, or it may be so protracted that surrounding circumstances may change, with unforeseen effects.

Moreover, large-scale conflicts are not simple or single affairs. They usually involve multiplicities of issues, parties and sub-parties. They are likely to involve conflicts and power struggles within as well as between parties (cf. earlier remarks on leaders and peace constituencies) and the stages of these internal conflicts may well not coincide with the stage the overall conflict has reached. The majority in a particular group may be ready for settlement with their original adversary, but facing a new threat from internal opposition. Even a 'peaceful' society will sustain a myriad of conflicts at any one time, all at different stages and following their own dynamics.

Nevertheless, despite all the cautions and disclaimers with which I surround the diagram in Figure 2.1 and its explanation, both here and elsewhere, in writing and in workshops, I have found that it bears some recognisable relationship to the experiences of those working with conflict and proves a useful tool for thinking about the stages and processes involved in conflict and its transformation. Even when the conflict in question begins in a way other than that depicted by the diagram, that very difference becomes the ground for fruitful discussion as to how conflict began. I have at various times developed more elaborate and broadly descriptive diagrams which plot different beginnings and less destructive trajectories (see, for example, the diagram in Figure 2.2), but the old original version, because of its relative simplicity, has proved more useful. It has been modified in small ways in response to new insights and suggestions from participants and further thinking of my own. Its use and usefulness will be often alluded to in the workshop accounts recorded later.

CONCLUSION

Conflict resolution offers a new approach to the problem of violence and how to escape its cycle. It has brought a new perspective, new insights and a new and developing body of theory. It has captured the imagination and energy of many, including individuals and organisations with some influence. It has received a degree of recognition in influential circles far beyond any given to nonviolence, whose extraordinary achievements seem to have been – perhaps studiously – ignored or forgotten. The support that conflict resolution has achieved, at all levels, offers a great opportunity for contributing to attempts to create new forms of violence containment and conflict transformation.

If this opportunity is to be grasped and the possibility of transforming conflicts of all kinds and at all stages is to have a chance of becoming a reality, the theory and practice of conflict transformation must respect the need and right of those most affected by violence to be active on their own behalf. Third parties have a role to play in the resolution of conflict, helping to build bridges of understanding and facilitating dialogue. A society with a constructive conflict culture will have its share of skilled and trained mediators, able to contribute in these ways to the handling of day to day conflicts, so that coercion and violence are prevented from becoming the normal means of dealing with differences. To focus exclusively on third-party intervention once violence has erupted is to ignore the primary task of enabling people to act constructively on their own behalf. Training is one form of conflict intervention which can go some way to addressing this need, but it is still often focused exclusively on third-party and bridge-building roles, ignoring the role of partisan actors and advocacy. The rich tradition of nonviolence, with its wealth of experience, can contribute to the correction of this deficiency.

Appendix 2.1
Complex Stages Diagram

STAGES AND PROCESSES IN INTRA-STATE AND REGIONAL
CONFLICTS

In any consideration of conflict and how to engage in or respond to
it, it is necessary to take into account its causes. Conflict in itself may
not be negative, but necessary. Violence, not conflict, is the problem.
The prevention or premature 'resolution' of conflict may mean in
practice the suppression of just aspirations: 'pacification' rather than
'peace-making'. At the same time, those wishing to enter into
conflict in the name of a just cause need to do so with some understanding of the likely cost to all concerned and of their current and
future possibilities in the light of the distribution of power.

Figure 2.2 describes the stages and processes which will usually
need to be passed through, if a situation characterised by the
oppression, exclusion or deprivation of one or more social groups,
with an extreme imbalance of power, or a situation of crisis or
turmoil, is to be transformed into one of genuine peace. (The words
contained in the rounded or oval shapes describe conflict stages,
while those contained in rectangles describe the actions or processes
by which new stages are reached.) The 'stages' are not in themselves
static. They have their own dynamics, and in practice they may
merge with one another. Neither are they likely to follow each other
in a clear and orderly sequence. It will often be a case of 'two steps
forward and one step back', or even vice versa; and frequently
processes need to be repeated, built on, reinforced by other processes,
in order to bring about substantial progress.

In practice the actions taken in conflict are often far from the
ideals of 'conflict transformation' and tend towards the perpetuation
of violence – direct, structural and cultural. Negative actions and
outcomes are indicated on the right-hand side of the upper part of
the diagram.

Large-scale conflicts are not simple or single affairs, but usually
involve multiplicities of issues, parties and sub-parties. They will in
all likelihood also involve conflicts and power struggles within as

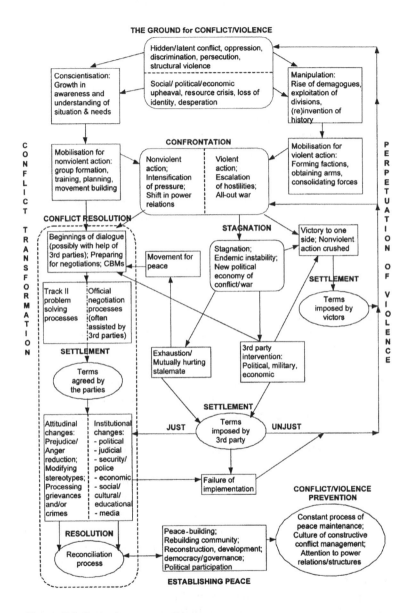

Figure 2.2 Complex stages diagram

well as between parties, and the stages of these internal conflicts may well not coincide with the stage the overall conflict has reached. Nevertheless this diagram may provide a useful framework for thinking about the stages of conflict and shifting power relations.

The diagram begins with two different types of situation. The first is one in which the oppression or exclusion of a given section of society is so complete that the conflict is hidden or latent, the oppressed group remaining passive in the face of extreme injustice or social violence (often maintained by physical violence, or the threat of it). They may remain passive because of tradition or lack of awareness, or if the power imbalance is such that they have no chance of being taken seriously in any demands or requests they might make. In order for this to change, some individual or group will need to begin to reflect upon, understand and articulate what is happening, and encourage others to do the same – the process described as 'conscientisation'. This process will, if it generates sufficient determination, lead to the formation of groups committed to change. Their first task will be to continue the process of reflection and analysis, formulating a common purpose and strategy, then developing organisationally as they begin to take action to build support, working, in the first place, for change through dialogue.

That is the ideal of active nonviolence or conflict transformation. What can happen, on the other hand, is that individuals or groups with their own vested interests or ambitions exploit the situation to their own advantage, whipping up hatred and generating explosive tension. (In practice, events can be expected to combine elements of the ideal and its opposite, since many different types of actors are involved.)

The second type of situation is one of deep upheaval or crisis, in which the socio-political structures and fabrics have been damaged or destroyed, and new forms need to emerge. Here again, social and political actors can work patiently to understand the situation and educate others with a view to a dialogical process of change, or they can exploit or manipulate for their own purposes, fomenting violence.

Even when actors are socially and idealistically motivated, some groups choose to use violence in their struggle. For others violence is not considered, or is not seen as a practical option. In the ideal of conflict transformation, there will be a strategic and principled choice for nonviolent action. However, as their power or visibility increases, as their voice begins to be heard, nonviolent movements may increasingly be seen by those in power as a threat, so that a stage of open confrontation becomes inevitable – a stage which may well involve repressive measures, including physical violence, on the part of those in power, even if the oppressed group have chosen to act nonviolently. During this stage of open conflict, the relationship in power between the opposing parties will change as a result of their ongoing confrontation (and other developments inside the parties or in the wider environment). It is possible that the nonviolent action will be met with such violence that the nonviolent movement is crushed, and terms and relationships for the foreseeable future imposed by those with the power to do so. Or the nonviolent actors may be drawn into violent conflict. Often, in practice, a conflict degenerates into violence through its own internal dynamics.

Violent conflict has a variety of possible outcomes, as suggested by the diagram: victory to one side, and terms imposed by them (which, although in theory could include and address the needs of the vanquished, are in practice likely to exclude or deny them); the forceful intervention of a powerful third party, leading to an imposed settlement (which could be wise and inclusive and pave the way for reconciliation, or could be unacceptable to one or more parties and lead to renewed violence or oppression), or exhaustion, a 'mutually hurting stalemate', or some other change in the course of the violent confrontation – such as the emergence of a movement for peace – leading to a search for dialogue.

If a previously oppressed or excluded group has increased its power or leverage sufficiently, whether violently or nonviolently, it can expect to be taken seriously as a partner in dialogue. When both or all parties perceive that they have more to lose than gain from continued confrontation, that is unlikely to produce an outcome that meets their needs, they will recognise the necessity of moving towards talks. At this stage it is possible to begin the processes grouped together and described as 'conflict resolution', in which communications are somehow restored and settlements reached.

This will not be a smooth process. Talks may break down, agreements may be broken, the conflict may flare up again. Facilitative non-partisan intervention can help, for instance, in the form of mediation, both in preparing the parties for negotiation (in problem-solving workshops, for example) and in the negotiations themselves. In these ways, and through face-to-face dialogue, some of the heat may be taken out of the situation, some more hope and trust generated, some of the prejudice dissipated, which in turn will facilitate the reaching of – and adherence to – agreements. Once these are in place, it may be possible to begin to deal with some of the remaining psychological damage, which the conflict and its causes have occasioned and to develop more positive relationships between the previously conflicting groups.

If the content of the settlement is not implemented and adhered to, this process will be undermined. Conversely, if psychological hostilities are inadequately diffused, implementation will break down. These two elements of resolution, the psychological and the substantive, are mutually dependent and need to go hand in hand. If both are adequately addressed, they will be further consolidated by projects for long-term peace-building and development for the well-being of the community as a whole, and by finding expression in the institutions and processes of society.

Social relationships are never static, and the final stage of 'peace' will need to be, in fact, a process (made up of a thousand processes) of maintaining awareness, of education, management of differences and adjustment and engagement at all levels, so that some new situation of oppression – or other major source of conflict – does not develop, and just and peaceful relationships are maintained.

3   Culture and Conflict Transformation

The theory and practice of conflict transformation constitute an endeavour to bring something new to human thinking and inter-action. Although in some measure the ideas they embrace are based on experience of what works (and what does not), the elaboration and presentation of these ideas and the activities based on them represent an attempt to effect quite fundamental changes in prevalent approaches to conflict and are therefore culturally challenging by nature. One of the criticisms levelled against what is more widely known as conflict resolution is that its Western origins make it untransferable, that its assumptions, diagnoses and prescriptions are so culturally formed and specific that they cannot but be misplaced and inappropriate in non-Western cultures, and that its promulgation amounts to cultural imperialism. Does that mean that all efforts to share and enrich, through cross-cultural work, the insights and experiences which are being accumulated for constructive approaches to conflict, the skills and ideas that are being developed, are misguided and counterproductive? These questions have never left me in my travels. They can be divided into three subsets: one is about the nature of the cultural differences in question; the second relates to the nature and importance of culture itself, and what, if anything, transcends it; the third is to do with the nature of the exchange on conflict resolution, and of the global relationships within which that exchange takes place.

Here I shall begin with a brief review of some of the issues often cited in relation to conflict and how it is dealt with in different cultures, moving on to the vexed question of the transferability of new models of mediation. After a discussion of the nature of culture, having asserted the need for change, I shall explore the application of universalist principles to conflict transformation, focusing on the concepts of human needs, violence and respect. And in the final portion of the chapter I shall look at the ways in which question of culture relate to the delivery and experience of workshops.

CATEGORIES OF CULTURAL DIFFERENCE RELEVANT TO THE TRANSFORMATION OF CONFLICT

Much has been written on the subject of conflict and culture (see Duryea's thorough and scholarly literature review and bibliography (1992)). Here I will simply list the most often cited cultural polarities relevant for conflict transformation, briefly evaluating their practical importance.

Individualism v. Collectivism

Broadly speaking, 'modern' cultures are characterised as being based on humanistic values which are focused on the life and dignity of individual human beings and 'traditional' cultures as being based on collective values and thinking. This dichotomy is seen as being linked to a Western emphasis on political rights and freedoms, as against the prioritising elsewhere of social obligations and economic justice. I have not, however, in practice, experienced any manifest conflict or contradiction between collectivist and individualist points of view – perhaps because social or collective action and inter-action is what conflict transformation is about (in contrast to stereotypical Western individualism), and because it is based on the assumption of individual responsibility *within* society, rather than detached from it.

However, through working with participants and colleagues from different continents, I have observed that there are different ways of organising and understanding collective identity. I have noted, for instance, the frequency with which Latin American participants have spoken of 'the people'. This concept carried powerful emotional overtones, associated with a sense of solidarity in the face of exclusion and repression. Europeans I have worked with have not referred to 'the people' in this way, identifying rather with particular cultural groups or political movements, and the notion of 'civil society'. I remember a Ugandan colleague telling me that for him neither 'civil society' nor 'the people' was a useful concept, and that in his culture one thought in terms of family or tribe. Particular individuals in whatever part of the world would, in any case, give different accounts of their own cultural realities. Nonetheless, it would seem that in all countries, if in differing degrees, people experience themselves both as individual beings and as members of groups.

Liberalism, Democracy and Egalitarianism v. Authoritarianism, Submission and Hierarchy

'Modern' societies aspire to liberalism and democracy, while 'traditional' societies place more emphasis on authority and submission. In the former, restrictions of liberty are to be justified, whereas in authoritarian societies and cultures, actions need permission. 'Modern' cultures tend to emphasise equality, 'traditional' societies hierarchy. These differences are certainly relevant for conflict transformation, which encourages the challenging of authority when it is oppressively structured or exercised, but again, in practice, I have not found a difference between activists of different cultures who, by definition, are those who are working for change. Conflict happens when power relations are challenged, so if conflict is present, someone's authority is already at issue. The acceptance of authority and status is, however, an issue in relation to 'facilitative mediation' (see below).

Issue Focus v. Relationship Focus

In 'modern' cultures, it is argued, people concern themselves only with the fulfilment of specific expectations, caring little for the religion or politics of those with whom they are not intimate, or about their relationship with them. In 'traditional' cultures, such matters are always important, so that the notion, often invoked in the conflict resolution field, of 'separating the people from the problem' makes little sense. I certainly reformulated my thinking and language on this matter after working in the Middle East, where the idea, so expressed, was clearly unhelpful – indeed provocative. However, I believe that it is also common in the West to personalise conflicts, and that it was in response to this tendency that the idea of focusing on the problem was formulated. It is also true in 'modern' cultures that positive personal relationships can contribute to the constructive handling of conflict (witness the difference made in East–West relations by the positive relationship between presidents Reagan and Gorbachev). At the same time, I have found that most people, regardless of culture, see sense in the idea of focusing on particular issues or behaviours in a conflict, which are capable of improvement, rather than attacking people (or groups or institutions) wholesale.

The Impact of Attitudes to Authority on Mediation

The role of mediators is often regarded as central to conflict resolution. It is argued that someone from a 'diffuse role culture' will not accept the 'Western' notion that the ideal mediator not only has no prior relationship to those in dispute but also exerts no influence over their choice of solution, being a process facilitator, not a solution adviser or adjudicator. In many cultures the mediator will usually be a wise and respected person, often holding social status on account of age and position, whose advice is looked for and whose task it is to advise and persuade. Again, I would argue that traditionally this is so also in the West, remaining the model for informal dispute resolution in families and organisations. It is the newfangled mediation of conflict resolution that has introduced the idea of the unknown and power-free mediator, who offers a service to those in trouble and whose role is simply to facilitate the joint deliberations of those in conflict. There may be less cultural difficulty in accepting such a concept in societies where the notions of individual equality and autonomy provide a counterbalance to hierarchical social patterns. And telling your secrets to strangers may be harder in some culture than others. However, in all cultures and at all levels, personal trust in the mediator will need to be established if his or her authority is to be willingly accepted rather than imposed, and informal mediation will continue to be done by individuals close to those in dispute.

To try to impose egalitarian processes in fundamentally hierarchical societies would be arrogant and foolish. To make them available for discussion and experimentation seems reasonable enough. However, enthusiasts for facilitative mediation should recognise that the use of traditional forms, in which a known and respected 'other' listens and advises, need not imply a lack of self-respect on the part of disputants, nor the abandonment of all autonomy. It indicates a choice to draw on another's wisdom and moral authority, rather than relying on one's own. It is, perhaps, useful here to distinguish between mediation and arbitration. It could be said that the traditional mediator combines the two functions, adjudicating and proposing while at the same time helping to restore relationships. At issue is the relative importance attached to autonomy on the one hand and interdependence on the other, and the role of status (and personal qualities) in social relations – which seems to be a matter for mutually enriching intercultural debate. The ideal of eventual willing ownership by the parties of the agreed solution to a conflict can still be met in authority-based forms of mediation or arbitration, given that the parties to the conflict willingly accept the process.

Analysis and Linearity v. Story-telling and Polychronology

Some practitioners and academics (for instance, Lederach, 1995 and Avruch, 1998) refer to the opposition between story telling and analytical tendencies in culture. Again, I am sceptical about the degree of difference that exists in reality, or rather about how far it is related to culture. The argument is that Western cultures favour analysis whereas 'traditional' cultures favour the more 'holistic' approach of story-telling. Maybe the academic world sets a premium on analysis, but from my experience as a mediator at home in the UK, I am convinced that story-telling is the first need and recourse of disputants, just as it is in the other contexts in which I have worked. Maybe the real cultural difference is between academics and others, or between theory and practice, and there is a need for theory to give due place to the role of partisan narrative. At the same time it is my experience that, in order to move forward from grievances about the past to agreement about the future, one must always reach the point of saying, 'So what does this amount to, and what's to be done?' – analytical questions.

Another categorisation of cultural difference is the distinction made between 'linear' and 'polychronic' ways of thinking: between examining issues step by logical step and discursively mixing different elements (Avruch, 1998). This difference, insofar as it is real, is of consequence for the stress placed in conflict resolution on analysis and, by implication, on corresponding order in the way issues are separated out and dealt with. However, although this may be the ideal of theoreticians or the preference of practitioners, reflecting some perceived Western tendency, in my experience it does not often happen in any culture – certainly not in my own. If any kind of agreement is to be reached, issues will need to be clarified to some degree, at least tacitly. In my own work as a mediator, I try to help those in dispute to identify different elements of what is at issue and to ensure that all those elements are covered in any agreement reached. But I do not expect the process by which this happens to be an orderly one, and would hope to coax it along rather than control it, accepting that dealing with emotional energies will be as important as the exercise of reason and order and that in reality the 'separate' issues are inextricably connected.

Material Motivation v. Identity Motivation

Max-Neef (1985) – outside the conflict resolution discipline, but relevant in his thinking – makes a distinction between the motivators or 'needs satisfiers' of the largely materialistic cultures of Western capitalism (and, till recently, of communism), with their emphasis on outer conditions, efficiency, rationality and action, and the emphasis he attributes to more 'traditional' cultures, on inner feelings and needs, and particularly on personal and collective identity and dignity, rather than material advantage. These different tendencies (like others) will not be confined within particular cultures, but may be more strongly accentuated in one culture than another. Both are represented within the conflict resolution field by 'realist' and what could be termed 'humanistic' approaches.

THE COMPLEXITY OF CULTURE

There can be no denying the importance of culture. It has been brought, unavoidably, into our consciousness by recent painful conflicts in relation to ethnicity and language, to the tension between the traditional, the modern and the postmodern, to North–South and East–West relations, and in particular in relation to the rising tension between Christendom and Islam. Awareness of culturally based assumptions and perceptions can contribute to self-awareness and understanding of others and increase our powers of choice. It can also illuminate our thinking about what needs to be addressed in conflict.

What Taylor (1994) refers to as 'the politics of recognition' of plural cultural identities within societies is widely accepted in the continents from which the idea of conflict resolution emanates. It is one of life's ironies that the perceived decadence of Western culture is epitomised by the relativist–pluralist approach of its philosophers. Yet it is that very pluralism that paves the way for the politics of recognition. Liberal philosopher Isaiah Berlin, who struggled to hold relativism at bay, nonetheless insisted that human identity was 'constituted by particular allegiances, cultural traditions and communal memberships, however complex and plural', and that there are no ultimate grounds for preferring one culture to another or invalidating one because it differs from another (Gray, 1995).

For liberal, egalitarian and pluralist cultures, the dilemma comes when the cultural form to be respected is an illiberal, hierarchical or monist one. How is illiberality to be respected by a liberal? But although a culture may be monist, it is unlikely to be monolithic. One of the critiques of multicultural politics is that members of a perceived cultural group may feel themselves to be trapped within its orthodoxies or identified by others with cultural attitudes and forms which they find inimical or abhorrent. To quote Avruch, 'the identities listed as social sources of culture are all potentially contestable and contested' (1998: 105). If I apply this thinking to myself, I notice that I identify myself in many different ways – human being, woman, English, Welsh, British, Quaker, critic of Quakers, post-Christian sceptical idealist, art lover, liberal-green-socialist, etc. – and that what is uppermost in my mind and in my self-presentation will vary according to circumstance. I realise, moreover, that I am both an insider and a dissident in many different communities, both real and 'imagined' (Anderson, 1983).

To generalise about cultures, whether of continents, countries or specific groups, is at the same time both necessary and nonsensical given the different discourses and influences which are embodied in the thinking of any human being, the differences between one individual and another within a family, and the differences between one country and another in a continent. Cultural generalisations, while they may be useful, must necessarily oversimplify, both in geographical terms and in relation to specific groups – which are, in the end, composed of varied individuals (my Western construction!). For instance, personal reticence is a relative matter. English people are probably, on average, far more reserved than their North American counterparts, though the two cultures have much in common. Teutonic and Latin styles in Europe differ considerably. And within whatever cultural grouping is chosen there will in any case be all kinds of variations and contradictions. Cultural generalisations can obscure important differences in power and interests between subgroups, as well as individuals, within a given group. Differences related to class or caste may cut across ethnic or other cultural differences. Although an awareness of cultural differences may be helpful in some ways, to generalise too broadly or to assume that all members of a given group will attach the same value to generalised cultural norms will prove counterproductive in terms of sensitivity. Whatever the chosen set of references, to frame things too readily in broad cultural terms can amount to prejudice and stereotyping –

a denial of individual values, feelings and needs. In their introduction to *Refusing Holy Orders: Women and Fundamentalism in Britain* (1992) Sahgal and Yuval-Davis, the editors, argue that multiculturalism imposes a religious identity on those who do not wish it, encourages religious fundamentalism, and bestows authority on male religious leaders at the expense of women within ethnic minorities. The promotion of gender equality and egalitarianism in general, and participatory processes for decision-making, can be expected to be (largely) anathema to those who are male and powerful – in whatever culture.

Taslima Nasrim (1997), writing in the *New Internationalist*, tells of her experience of being a 'disobedient woman' in Bangladesh, and concludes that the moral standards of any society are fluid and relate to its economic situation, political structure, religious influences and system of education. Her argument for personal moral responsibility and the supremacy of individual moral understanding and conscience comes close, it seems, to the Western valuing of 'autonomy'. It also comes close to my idea of respect as a fundamental and universal value. Duryea (1992: 18) cites Ishimaya as arguing for the importance of 'transcultural individual identity' as suggested by Jung, Maslow, Erickson and others, as the basis for a deeper understanding of life and basis for human interaction, 'compared to a culture-bound, unidimensional self-identity'. Duryea also cites Fisher and Long and their study findings which suggest that matching mediator and client ethnicity in Alternative Dispute Resolution (one manifestation of conflict resolution, which began in the US as an alternative to litigation) is not as important as other things, particularly gender and general life experience.

Cultures are not static, any more than they are monolithic or bounded. They are flowing, overlapping, kaleidoscopic, constantly moving and changing. The magnitude and pace of wide-scale cultural change may vary from one society to another, but even in the most traditional of societies there will be constant, if small-scale or subtle, change, because human beings are not simply functions of their culture but actors who make choices. They are in turn formed by culture and formers of it, both perpetuating and changing it. Many societies in the twentieth century have been torn apart by the pace of change – political, economic and cultural. Whether or not this in itself is really something new, I am not qualified to say, but I see some of the new separatist/nationalist energies, with their re-invention of history and 'traditional' culture (often accompanied by ethnocentrism and indeed racism) as a reaction to political, economic and cultural earthquakes which together constitute every kind of existential threat to those caught up in them. Those affected can see themselves (and be seen as) the victims of history, or as its enactors, or both. They will have many choices to make, collectively and individually, including cultural choices. Processes for conflict transformation are based (whether explicitly or tacitly) on the assumption that human beings, of whatever culture and in whatever circumstances, retain some capacity for choice, however limited that choice may be and whatever events are in train. The processes are designed to facilitate and maximise choice – in the first place by bringing its existence into awareness.

SHIFTING FOUNDATIONS AND THE NEED FOR CHANGE

The shrinking of our planet has confronted us in a new way with an old problem: that human differences, individual and collective, and human needs for identity, security and control together make conflict inevitable. The technological revolution, which brings us closer together and makes us more than ever interdependent, has done nothing to improve our skills in dealing with each other, only increased our capacities for destruction. Moreover, colonial history casts its shadow over current perceptions and is reflected in power relations which remain cruelly asymmetrical, both globally and within societies. Issues of difference, of culture and identity become tangled with issues of justice and injustice, and the resulting conflicts are very often asymmetrical in terms of power. While new tyrannies replace the old at the local level, the overall distribution of economic and political power is largely undisturbed, though its continuation is morally and probably, in the long run, practically indefensible.

Those who work for the transformation of conflict believe that it will always be present in human society and that what is both necessary and possible is to find ways of dealing with it constructively. That means, among other things, reducing and as far as possible eliminating the violence associated with it. Given the number of violent conflicts that scarred the last century, and the suffering, displacement, famine and destruction they caused, this seems an important project. It is apparent that current mainstream approaches, which are based on coercive power, do not hold the key. The story of wars throughout the ages has been about the struggle for domination rather than coexistence (Eisler, 1990), a story of the untold suffering of combatants and non-combatants alike, in which outcomes are decided by a combination of luck and power, a story in which even victories involve huge losses, misery and destruction, and in which the seeds for new wars are sown.

UNIVERSALISM: BASIC HUMAN NEEDS, VIOLENCE AND RESPECT

The development of the field known as conflict resolution (or transformation) represents one attempt (though certainly not the first) to change cultural approaches to conflict and the exercise of power: to find alternatives to violence as the agent of change; to replace dominatory processes with co-operative ones; to work for inclusive solutions to conflicts, addressing the needs of all parties, rather than seeking the victory of one side over another. Such a project of cultural change must· be based on some foundational values and assumptions, though these are not often made explicit in the literature. Isaiah Berlin asserted that 'there exist central features of our experience that are invariant and omnipresent' (1978: 164–5). The concept of basic and universal human needs is predicated on this assumption and has been given a central place in conflict resolution theory by John Burton (for example, 1987, 1990) and many others. According to this thinking, conflicts cannot be properly resolved – that is, agreements satisfactory to all sides (and therefore durable) cannot be found – unless the fundamental needs of the parties are identified and met: needs such as security, identity, recognition and participation (or, according to Galtung (1990), the needs of survival, well-being, identity and meaning, and freedom). Jay Rothman, for instance, argues for processes in which parties to a conflict are enabled to see that 'adversaries, like the self, are deeply motivated by shared, human concerns and that, unless these are fulfilled, violence will be perpetuated' (Rothman, 1992: 62). The 'satisfiers' or ways of meeting different fundamental needs will certainly be culturally influenced (Max-Neef, 1985), and the connections between needs and demands in conflict may therefore at times be hard to understand, but this does not fatally undermine the usefulness of a needs-based approach.

The process of co-operative problem-solving advocated by Rothman and others respects the dignity of all parties, reframing the conflict in terms of a joint attempt to find a way forward, in place of continuing mutual assault. Thus it becomes possible for the conflict to be viewed inclusively and seen as a shared problem rather than in terms of opposition. Such shifts in perception are seen as possible if one assumes not only the existence of universal human needs but a human capacity for empathy for making the imaginative leap from what is needed for our own well-being to an understanding of others' needs. Such a capacity is implied by the concepts of 'natural law' and the 'golden rule' contained in all major world religions (however persistently flouted by some of their adherents): 'Do as you would be done by.'

In understanding the needs of others we are, however, confronted once more by the question of culture and difference. Manfred Max-Neef, contributing to discussion of *The New Economic Agenda*, supports the notion of universal human needs, and gives his own list of basic needs (1985: 147):

> These fundamental needs are in our opinion the same for every human being in every culture and in any period of history. They are the needs for permanence or subsistence; for protection; for affection or love; for understanding; for participation; the need for leisure; for creation; for identity; and for freedom.

He categorises these needs as 'those of having and those of being', which 'not only can but must be met simultaneously', and suggests that in the West we have concentrated on having at the expense of being, and that 'being' needs are as important as material ones.

Max-Neef's insistence on 'being' or psychological needs is supported by James Scott's powerful work, *Domination and the Arts of Resistance: Hidden Transcripts*. Scott lays his emphasis on the psychological aspects of oppression (1990: 111–12):

> the social experience of indignities, control, submission, humiliation, forced deference, and punishment ... the pattern of personal humiliations that characterize that exploitation.

'Dignity' and 'indignities' are words Scott uses repeatedly to indicate the need for recognition and respect. If material needs alone are the focus of negotiation in conflict and no place is made for what is at issue at other levels, the conflict is unlikely to be resolved in any satisfactory way. And if we assume, from our own cultural viewpoint, that we know what matters most to people and fail to listen attentively to what they say themselves (fail also to try to listen to what is not said), we are likely to miss the real understanding that is needed.

It is argued by Rothman (above) and others that basic human needs are 'non-reducible' and therefore non-negotiable. In the sense that people cannot be argued out of them, that is true. At the same time, it is part of the human condition that no one's need for security or recognition – or anything else – is ever met in an absolute way. We have to make do with *relative* security, a *degree* of participation, and so on. And as Berlin (above) insists, one social good has to be weighed and balanced against another. What I believe is necessary in negotiation is that the fundamental human needs at stake are recognised and acknowledged (though not necessarily explicitly), and receive what is felt to be an adequate response in any proposed solution. The process of negotiation itself should meet the human needs of security, participation and recognition. Those involved need to be aware that cultural and other factors may mean that the needs 'satisfiers' for one individual or group in a given situation (including the process itself) may be different from the things which would satisfy the needs of another individual or group in similar circumstances.

As already indicated in Chapter 1, the concept of violence is not limited to that of physical assault. Violence as understood here can take many forms and is defined by Johan Galtung (1990) as 'avoidable insults to basic human needs'. It is to act in ways that deny the humanity of another or dehumanises them. According to Freire (1972: 20):

> Dehumanization, which marks not only those whose humanity has been stolen, but also those who have stolen it, is a distortion of the vocation of becoming more fully human.

If violence is understood in this way, then respect may be seen as its opposite: the recognition of the true nature and potential of others, the acknowledgement of their reality and needs, the will to make space for them and honour them.

Respect between human beings involves not only the recognition of individuality and interdependence but the affirmation of some kind of fundamental equality, however hard that equality may be to define. 'All human beings are born free and equal in dignity and rights.' So runs the UN declaration; but in many ways individuals are clearly unequal – in size, ability, status, for instance – and some are enslaved from the moment of their birth, their fundamental rights denied. So in what does this 'equality' consist? Antoine de Saint-Exupery, writing (in the male language of his time) of his rescue from death in the desert, addresses (rhetorically) his desert rescuer (1995: 102):

> as for you, our saviour, Bedouin of Libya ... I will never be able to remember your face. You are Man, and you appear to me with the face of all men together. You have never set eyes on us, yet you have recognised us. You are our beloved brother. And I in my turn will recognise you in all men.

Recognition of this kind discounts difference, looking beyond it to one common identity. The most fundamental form of respect, which is a response to humanity itself, is a recognition that any other human being has an equal stake in life and the same fundamental needs as oneself – a recognition that we are all members of one species, products of the same evolutionary process and occupying one ecological niche. We share our capacity to dream and our knowledge of mortality, our powers of choice and our ultimate helplessness. We need each other's solidarity, both practically and psychologically. In acknowledging all this we respect what is common.

Respect, however, also recognises and honours what is different. In the words of Gerard Manley Hopkins (1953: 51):

> Each mortal thing does one thing and the same:
> Deals out that being indoors each one dwells;
> Selves – goes itself; *myself* it speaks and spells,
> Crying *what I do is me: for that I came.*

(Compare Gandhi's idea of 'self-actualisation', referred to in Chapter 2.)

Rockefeller (1994: 87) brings culture into the equation, arguing that respect for the universal and the particular potential of every human being involves:

> respect for the intrinsic value of the different cultural forms in and through which individuals actualise their humanity and express their unique personalities.

The notion of human rights is based on what is common, but is applied through the recognition of individual autonomy and difference, and the right to association. It calls us to respect all our forms of identity, individual, collective and universal.

Such fundamental respect is unconditional and has nothing to do with the kind of social regard which is afforded to those with particular gifts, possessions, power or status. Neither is it dependent on the behaviour of recipients, but is to be maintained in the face of all kinds of provocations, disappointments and outrages. On the basis of such fundamental, unconditional respect, secondary, conditional, moral respect will be attached to those whose behaviour is respectful of the being and needs of others. Respectfulness engenders respect and honours the being of the giver as well as that of the recipient. Just as, in Freire's thinking, those who dehumanise others dehumanise themselves, respecting others provides the grounds for self-respect and is the expression of it. The lack of self-respect which leads to disrespect of others can be remedied only by the experience of that basic, unconditional respect or regard which persists in spite of all kinds of provocation and has the potential to transform. A society which embodied, or at least aspired to, such unconditional respect for humanity, in general and in particular, would constitute a truly civilised society.

This was the concept with which I began my doctoral research, more than five years ago (Francis, 1998), as a practitioner in the field of conflict transformation. I wanted to discover whether respect was a value which could provide a positive counterpart to violence and a common foundation for human relationships and for conflict transformation. I had read in Duryea (1992: 18) that it is 'mentioned by many writers as a key element that transcends culture'. It has since been my experience of working with people from every continent that this is indeed so. I was, to begin with, a little afraid that differences of language and culture might render impossible any meaningful discussion of the nature and application of respect. In practice I experienced that it had mutually intelligible meaning in all the groups in which I worked, with participants from all continents. It proved a vehicle for animated dialogue, both about the concept itself and about the experiences and feelings it represents and defines. It was never called into question as a value, though the emphasis it was given and the manner of its application might be strongly at issue. It was an idea that seemed to get to the heart of what conflict and its transformation are about.

CHALLENGING CULTURAL VIOLENCE

Ideas matter. Behind our ways of behaving and our social structures lie the patterns of belief, thought and expression which we call culture, and values are at culture's heart. In Chapter 1 I referred to Johan Galtung's concept of 'cultural violence', which 'makes direct and structural violence look, even feel, right – or at least not wrong'. If I look at the cultural fabric of Britain, for example, I find no shortage of examples of cultural violence and its effects. It is embodied in mainstream political language – the language of 'deterrence'; of the 'judicious', 'necessary' or 'firm' use of 'force', and of 'economic necessity' and the 'protection of national interests'. These euphemisms embody the culturally entrenched position that killing people, destroying homes and infrastructures, laying waste the land and contaminating earth, sea and sky, allowing sickness and starvation which are preventable – all constitute justifiable conduct, whose enactment or threat is embedded in the norms of international behaviour and relationships.

Though Western governments are, for historic and economic reasons, in a position to extend their military and economic bullying to faraway places, those who live in western Europe and North America are by no means alone in the cultural enshrinement, promotion and justification of violence, both structural and direct. Libertarian, individualist, capitalist cultures, founded on imperial exploitation, are arguably no more or less culturally violent than authoritarian and repressive societies and systems of the more 'communitarian' and 'traditional' kind. Oppression of one sex, class or caste, or of one ethnic or religious group by another, is not a Western prerogative. To justify such oppression in the name of culture remains a matter of cultural violence, whatever the culture in question.

Within the thought frame of conflict transformation, parity of esteem – equal respect for all persons in a process – is fundamental, and implicitly, if not explicitly, challenges the hierarchical valuing of the being and needs of different categories of people. Also central to conflict transformation is its challenge to the role which violence plays in most cultures as the chosen means and arbiter of conflict in certain contexts. The cultural acceptance of violence, which I see as a problem in my own society, is seen as such by colleagues all over the world. Yet in most cultures there are other strands which encourage respect for others and for life. So Gandhi drew on the Hindu value of *ahimsa*, and absolute respect for the human person is fundamental to Gandhian nonviolence.

I start, like Taylor (1994), from the presumption until proved otherwise of cultural equality. It makes sense to me to assume that all cultures have some elements which are positive and some which are negative in their effect upon the way conflict is handled. Some cultural attributes will be both positive and negative in their effect. The value given to patience in a certain culture may help people to cope with relentlessly harsh conditions and at the same time encourage passivity in the face of gross oppression. The strong social cohesion within a tribe or village, the value given to belonging, may also find expression in violent hostility to outsiders. Adherence to tradition about different social roles may bring stability. But it may also involve the appeasement of those in authority when they abuse their power, and the justification of discrimination against people on account of their age, gender or social position. The value of respect can provide a common reference point for evaluating cultural norms. It can also provide the basis for a cultural shift which weakens the structures of oppressive relationships and the tendency to resort to violence in conflict, encouraging us to find more constructive ways of thinking about it and dealing with it.

Inter-group conflict often involves cultural differences of some kind. If the dialogical processes proposed by conflict transformation are to be of use in such circumstances, then cross-cultural dialogue must also be possible, and I believe it is. I have no doubt that important cultural differences exist and can be described in a variety of ways. I am aware also that my own perceptions are culturally influenced, if not determined, and that this awareness is important. At the same time, I am convinced that attitude, language and behaviour, or respect, can provide a bridge for crossing the many and varied divisions, cultural and otherwise, between individuals and groups, providing a common reference point for perceptions and aspirations.

CULTURE AND WORKSHOPS

Workshops often constitute microcosms of cultural interaction. Facilitators of *dialogue workshops* will be able to be effective only if they manage to put the participants at ease – with themselves and eventually with each other. Feeling respected is often a precondition for being willing to communicate constructively, and the form of the dialogue which is to be attempted needs to be agreed by all who take part in it. In my experience workshop participants are willing, if adequately informed, to take part in processes which are new to them and to take responsibility for the outcome. Knowledge of cultural norms within the participant group will help the facilitators to read what is happening and behave appropriately themselves, but close attention to the actual dynamics of the interactions that take place may be more important than cultural generalisations. (In Chapter 8 I shall discuss issues of good practice for outside facilitators.) The higher participants are positioned in their own social or political scale, the more sensitive they are likely to be to the ways in which respect is conveyed. To begin in an acceptable setting and with expected formalities may be important in paving the way for less conventional processes in which 'parity of esteem' becomes the norm.

Concerns about cultural awareness and the transfer of ideas and methods are perhaps mostly directed towards *training workshops*, where this transfer is seen to be direct. The word 'training' is perhaps unfortunate, suggesting the one-sided delivery of knowledge and transmission of skills. I see my role as a trainer as being to devise and facilitate mutually educative processes. For me, therefore, training has participants rather than recipients. Through the processes offered, participants have the opportunity to clarify and deepen their understanding of their own experience, enlarging it through exposure to the experiences and understandings of others, and developing skills for acting on their own values in pursuit of their own goals. Thus, whereas 'training' might be taken to mean 'formation' (to use the French term) by a third party, in my understanding it provides the opportunity for self-development and mutual education. And whereas training could be understood as limiting behaviour to desired patterns, I would see it rather as adding to its possibilities, as widening rather than narrowing choice, enhancing participants' capacity to think, analyse, observe, evaluate and make decisions, both over time and in a given moment, and their confidence that they have these capacities.

The values of conflict transformation training are those which inform conflict transformation more generally – those of respect, parity of esteem, participation, consensus and inclusiveness. I hope, then, to avoid cultural or any other form of imposition by using a broadly elicitive approach (as against what Freire (1972) calls the 'banking' approach) to the learning process, building it on the existing wisdom and experience of participants (see Lederach, 1995; Fisher, 1995 and 1997; Babbit 1997). I do not wish to suggest that facilitators with an elicitive style make no input. The very notion of conflict transformation, by whatever name, the structure of the agenda (or the method of its construction, if it is done with participants), the choice of content elements, the workshop style and ethos, the underlying goals and values on which it is based, will reflect the theories, norms and values of the trainers and the organ-isation for which they work. Such a style does, however, provide the maximum opportunity for participants to begin where *they* are, to base their thinking on their own understanding, to evaluate, select and reject new ideas they are offered, and to integrate them, if they choose, with their own existing ways of approaching conflict (see Fisher, 1995: 13). Simply to make one's own implicit knowledge explicit can be a powerful experience. (I remember the words of one participant: 'The things we've been learning have been things we already knew but couldn't use because something in our thinking was stopping us.')

I do not wish to overstate the power of an elicitive training style to leapfrog over all cultural divides. Inclusive, egalitarian, informal and participatory processes are not culturally neutral. They embody the counter-cultural values of conflict transformation. Workshops of this kind come as a culture shock to participants from almost anywhere (including western Europe), when encountered for the first time. They create, in themselves, a temporary and intensely experienced culture in which participants speak and act in ways they would, perhaps, rarely if ever do in the outside world. For instance, the degree of frankness which allows for mutual and public self-evaluation is likely to be counter-cultural in itself, which does not render the process impossible, but can make it a very sensitive matter. For those who understand learning as being delivered by experts in the form of lectures, to be invited to work in small groups, experiment with ideas, participate in role-plays and give equal space to colleagues who outside of the workshop are not their hierarchical equals, can be seen as something of an insult – as not being taken seriously as students or as people of a certain status.

The informality and intimacy of such processes can be quite chal-lenging and uncomfortable for some, though participants can, in theory, choose their level of engagement. The fact that workshops create their own temporary culture is part of their power. The intensity of the experience means that the learning is also intense. This is also a potential weakness, since there is a danger that the behaviours and ideas explored and practised in this temporary culture may not be easily transferable. There may be elements of it, and counter-cultural ideas it has generated, that participants wish to incorporate into their lives outside, but the difficulties should be carefully considered with participants, and not be underestimated. Lederach (1995) suggests that we need to progress beyond a concern for cultural sensitivity to a recognition that the solutions to conflict taking place within a given culture must come from the resources of that culture, and he and others have written of traditional approaches to conflict-handling in different cultures. If any culture has useful models to offer, we need them. But to rely only on tradition would suggest too static a view of culture, for all the reasons given above.

Given the human misery that our different cultures have produced, it seems reasonable to re-examine traditions and assumptions, to engage in culture critique and acknowledge the need for cross-fertilisation and invention. If, as Freire (1972) suggests, the heart of education is 'conscientisation', or bringing into awareness, the most fundamental form of education, and arguably the most powerful or empowering, is to bring into awareness the cultural patterns of thought and action which, unless we recognise them and choose otherwise, form our lives.

EVALUATION OF CULTURE AS PART OF THE WORKSHOP PROCESS

'Conscientisation' (Freire 1972) means bringing into awareness and evaluating the different structures and influences that affect one's life. Cultural reflection is part of that process. As a colleague from Niger once remarked, 'There is something liberating in discovering the limitations of one's own culture. People are often oppressed by something in their culture but can't name it, and it's a liberation when they do.' I have found that this view of education as a means of liberation is one shared by people in every continent, who see it as the route both to self-respect and to social power and responsibility.

Often debate about culture happens spontaneously during workshops. For example, in one multicultural workshop I was facilitating, an African woman defined the custom of female genital mutilation as cultural violence against women. An African man countered that maintaining such traditions was an important way of resisting the cultural imperialism of the West. The woman retorted that no one should justify violence against women in the name of culture or anti-colonialism and that culture was not something fixed, but constantly re-created. In another pan-African workshop, the women chose to use conflict transformation analysis to address the injustice of the traditional exclusion of women from inheritance rights. At an international trainers' gathering, a fellow trainer from India described how she works with women to help them recognise their own dignity and rights and to stand up for them in the face of prevalent customs and practices. The 'women's movement' seems to be alive and kicking in many, if not all, cultures. There are also women who disapprove of it, and among men there is widespread fear and resistance, since the cultural challenge it represents is experienced as an attack on their identity.

Oppressive restrictions on young people represent another form of culturally sanctioned discrimination which I have heard criticised by colleagues from the Indian subcontinent, reflecting on social patterns in their own countries. As they see it, the opinions and needs of younger family members are not respected. And the wider culture of reticence in expressing opposition or disagreement compounds the problems of the young, who are silenced both by this general cultural norm and by the special limitations associated with their status in society as young people. The culture of silence or reticence referred to here and earlier has been discussed in several multicultural workshops that I have facilitated, by participants from cultures where such reticence is highly valued. These discussions, which (ironically and perhaps significantly) were characterised by open disagreement, demonstrated the possibility of stepping outside one's own cultural constraints. They also constituted a healthy challenge to the rather overworked Western notion of assertiveness. I give these examples not as substantial evidence of some 'truth' about particular societies, but to indicate that some members of those societies take a critical viewpoint of certain cultural tendencies they identify in them.

The difference between 'covert' and 'open' communication patterns has at times been a topic of discussion in workshops I have facilitated and is significant for the actual process of workshops. The idea that individuals involved in conflict should speak openly of their feelings and needs sits uneasily with the value placed by (for instance) Asian society at large on personal reticence and communal harmony. In inter-group relations this reticence may be overcome by the forms and demands of, for example, negotiation procedures, but in interpersonal relations within a multicultural workshop group, or when dialogue between Asians is mediated by a non-Asian who is unaware of the constraints, the difference may present problems for all concerned. For the workshop facilitator, it may mean that feedback cannot be taken at face value, in that dissatisfaction or discomfort may be covered by politeness. What is seen in the West as helpful frankness may be seen in the East as rudeness and as a disregard for the dignity of all concerned.

In practice I have found that workshop participants from, for instance, India, Japan and Korea have spoken quite openly within the workshop context and have had interesting observations to make on the advantages and disadvantages of their own social norms. Their observations have helped other participants to be more aware of the tone and impact of their own communications and the range of responses they may engender. Nonetheless, the potential implications for cross-cultural communication are real and will be illustrated in Part II. The relationship between the values of outspokenness (overt communication) and heterogeneity is close. Both are related to an emphasis on the rights of the individual to self-expression, as against the demands of conformity to a tacitly agreed identity and behavioural norms. Again, however, I have experienced that Western and Eastern workshop participants are able, at least, to communicate about these different value weightings in their cultures, and also differ between themselves on the importance they place on different values.

Since challenging culture presupposes moral assumptions that provide a yardstick for its evaluation, I often introduce specific opportunities for reflection on values, which will be discussed in Chapter 4. The arguments that arise and the dilemmas that surface have recurrent elements. Wherever participants live and whatever their culture, they are confronted by the problem of weighing and balancing competing values and needs. This convinces me that some aspects of human experience and response do indeed cut across cultural divides. One such aspect is the difficulty of living out competing values (Berlin, 1998 and elsewhere). It follows that we must and do take responsibility and make choices, debating these things both internally (see Billig, 1987) and socially.

I am not attempting to argue that there are no cultural differences about values, only that there appears to be also some common ground which provides the basis for a consideration of constructive approaches to conflict and that, whatever our values, it is a difficult and infinitely complex business to try and think about them, let alone live them. I am also indicating that participants in processes I have facilitated have been open and indeed eager to explore these dilemmas and have cared passionately about them. Through identifying and discussing the values they hold most dear, workshop participants find a reference point for their own cultural evaluation. And through these processes I have been involved in constant debate about my own assumptions, convictions and allegiances, both within myself and with participants. I have been changed by these interactions, just as I have seen others change.

A NOTE ON LANGUAGE

Language both embodies and forms culture. It is also a powerful element of identity and an instrument of power. Linguistic domination is frequently an issue in conflict. Those who speak dominant languages should do so with sensitivity and do all they can to share their power of communication. The management of language in workshops will be discussed in Chapter 4.

CONCLUSION

From my own experience I have concluded that the thinking and processes of conflict transformation do have relevance for different societies and cultures, even if that relevance is culturally revolutionary. While cultural identity is frequently one of the things at issue in social and indeed international conflict, dealing constructively with difference is what conflict transformation is all about. Cultural differences become seriously obstructive and a source or focus of deep misunderstanding and irritation when those differences are associated with substantial power disparities or with an overarching sense of grievance (Francis, 2000b). Then culture may become a vehicle for separation and polarisation and the fomenting of hostility. In such circumstances, the counter-culture of conflict transformation may provide a bridge that can transcend the cultural element of the conflict and offer the chance to use and develop a new common language for dealing with it.

What is, perhaps, most at issue in the culture debate is whether people from North America and Europe can be of use, as consultants, facilitators, mediators and trainers, in other parts of the world. The field of conflict resolution has been developed predominantly in the West, particularly in the US, although it has, under a variety of titles (conflict management/handling, conflict transformation, reconciliation) some notable proponents from other parts of the world. I believe that the problem here again is power and its abuse, past and present – in the past through colonial wars and occupations, in the present through economic exploitation and military domination. Disrespect for other peoples and their cultures is part and parcel of these power relations.

It is this fundamental and massive asymmetry which makes on-the-level relationships between 'the West and the rest' difficult in any context, from the interpersonal sphere to the governmental, and packs the culture debate with dynamite. Culture becomes a focus for resentment about colonialism, past and present, and stands for dignity and identity in the face of humiliation and denial. From this viewpoint, the Western concept of conflict resolution (so much at odds with the actual conduct of Western governments in international relations) must seem a very suspect gift from the powerful to the disempowered. It is not surprising that it is felt by some to be patronising and inappropriate, nor that it is seen as a potential Trojan horse – a way of smuggling in the means of pacification, of dissipating the energies and the will of the disempowered for fighting for their rights, while the West continues to maintain its power in the way it acquired it, through massive military violence, either threatened or actual. According to Abu-Nimer (1997: 31) the 'core dilemma' with which the conflict resolution field is faced is whether it can meet the needs of 'deprived groups who lack the basic needs' rather than serving the interests of those who benefit from the status quo through its 'no blame' (and therefore 'no justice') approach and its emphasis on 'conflict prevention'.

My conclusion therefore is that those of us who come from the West, if we are to contribute to conflict transformation globally as well as locally, must cultivate our own humility, and be aware not only of our cultural roots and bias, but of any unconscious neo-colonialism or paternalism in the way we approach our work, putting ourselves at the service of others rather than seeking to impose our views on them. We must ensure that our approach and activities are supportive rather than directive and that they are indeed wanted rather than inflicted. We must be aware of the deficiencies of our own societies, cultures and governments, while open and honest in speaking our values. And we must ensure that our understanding of conflict transformation includes the value of justice and upholds the need for change, drawing on the insights and experiences of Mohandas Gandhi and of nonviolent activists in Latin America, learning from those who are drawing on tradition to pioneer new approaches to reconciliation in Africa, and from many others in the South who can teach us about justice and nonviolence in dealing with conflict.

I do not wish to minimise the difficulty of any radical re-formation of human culture and institutions, or to be what Avruch (1998) describes as an 'overzealous conflict resolver', carried away by 'the natural optimism and horizonless possibilities of a young field'. (As Rouhana (1995) suggests, to claim to be a practitioner of conflict resolution is in any case a little immodest. To hope to make some small contribution to the transformation of conflict would be more appropriate.) But if we resign ourselves to pathological relationships, to the oppression of some human beings by others and to the inevitability of war and destruction, investing culture with the right and the power to control our lives, I believe we degrade ourselves, robbing ourselves of moral choice and responsibility and condemning ourselves and others to witness and suffer atrocities that are committed by human agency. Conflict transformation implies the acceptance of personal and collective responsibility. It also means culture transformation for all of us.

Part II

From Theory to Practice: Training and Dialogue

I have, in Part I, offered my own world view and rationale for conflict transformation, discussed the range of theory I believe to be necessary to it, and indicated some of the questions and sensitivities related to power and culture of which practitioners should be aware.

In this second part of the book, I want to turn to practice. I shall focus on workshops, the most used method for changing attitudes, building peace constituencies and developing capacities for conflict transformation. They are also the most-used means of constructive encounter between opposing parties in regions of conflict. In Chapter 4, I will explain, in general terms but at the same time in some detail, the aims, content and methodology of workshops, which are based on the ideas set out in Part I. The other chapters in Part II will bring these different elements to life. To do this I shall use as illustrations accounts of three different training workshops, and a description and discussion of an extended series of workshops focused on dialogue. (The accounts vary in length and style, since they were written in different circumstances, while the workshops they described and the issues they raised were still vivid in my mind, and I do not want to edit them for fear of losing their immediacy.) Each of these chapters will conclude with some reflections.

The first account, in Chapter 5, records the day-to-day intricacies of an introductory training workshop on conflict transformation involving participants from every continent. This example illustrates in practice the broad theory of conflict transformation discussed in Chapter 2 and the training approach and methods I am about to describe. It also demonstrates some of the cultural issues already alluded to, with their associated power dynamics, particularly in relation to gender. Chapter 6 records a workshop held in Africa for women trainers, and illustrates, sometimes painfully, the tensions involved in North–South relations, as well as questions about different ways of addressing gender issues. Chapter 7 contains the account of a training workshop held with a group of women from different parts of the former Yugoslavia, while the 1991–95 war there was in progress. It shows the way in which the dynamics of that conflict made themselves felt within the participant group, and how learning and bridge-building were combined. Chapter 8 is an account of a succession of workshops, which began as an aid to dialogue about the future of Kosovo/a, before the NATO bombing and international occupation. It traces the changing shape and nature of the process in response to the very different situation that that intervention brought about.

Before returning in Part III to the broader discussion of support for conflict transformation, I will summarise, in Chapter 9, my thinking on some of the issues raised in Chapters 5–8 which I have found important in workshop facilitation.

4   Workshops: Aims, Content and Methodology

Since the many individual practitioners (and the organisations they represent) who arrange and facilitate workshops will have different ideas and approaches, I cannot speak for them all, but only from my own experience and understanding. I would hope, however, that the approach, content and methods I describe will be recognisable or at least seem reasonable to other facilitators and trainers. Usually I work in a two-person facilitation team, and change, adapt and expand my own ways of doing things in response to the ideas and styles of my colleagues. However, I shall talk here about what 'I' do because I can generalise only about myself.

As I suggested in Chapter 1, workshops may be roughly divided into those focused on training and those focused on dialogue – they could be distinguished as workshops *about* conflict transformation and workshops *for* conflict transformation. However, the distinction is often blurred. For instance, the two types overlap when training is used as a medium for dialogue or a route into it, and in any case participants in dialogue learn a great deal about conflict transformation and its methods. Similarly, workshops held with groups intending to take action together, even when they are described as training workshops, are part of the 'conscientisation' process, which is an essential element in conflict transformation. So although I shall write about training and dialogue under separate headings, the reader should bear the overlaps and hybrids constantly in mind. I shall start with an extended discussion of training workshops, devoting the larger part of the chapter to them. I shall do so not because they are more important (though they are certainly more numerous) but because in so doing I shall have laid out the pieces of a puzzle from which I can then, relatively quickly, construct a picture of dialogue workshops.

I have structured the chapter under the headings of aims, content, design, methods and facilitation, with some final notes on practicalities. I begin here by elaborating further on the aims of training workshops as set out in Chapter 1.

AIMS OF TRAINING WORKSHOPS

It is important from the outset to have a common understanding of a workshop's aims, decided by local organisers and international partners (if any) on the basis of their assessment of what is needed and made known in advance to participants.

The clearest purpose of training workshops is, generally speaking, to help participants to find new ways of understanding conflict and constructive approaches to it, and to support them in identifying and developing skills and capacities for engaging in it and coping with it in positive ways. The skills developed can be divided into three broad categories: (i) the internal skills of self-awareness and reflexivity; (ii) the external skills for effective personal behaviour in conflict; and (iii) the analytical, organisational and strategic skills for effective group action. The broad aim of training workshops is to contribute to the empowerment of those affected by conflict, supporting them in acting to reduce the violence with which it is associated. Whether they be activists from popular movements and NGOs, public officials, diplomats, journalists, religious leaders, or a mix of concerned individuals, the goal is to give them an opportunity to review and develop their options to become agents for the promotion of human rights, initiators of dialogue, creators of alternatives, builders of peace constituencies, and constructive negotiators.

Although these workshops may take place under the title of training, and despite the fact that I have summarised their purposes in terms of understanding and skill, the support they can offer is in practice much wider and, perhaps, deeper than that. Groups of activists (or would-be activists) in violently conflictual circumstances often find themselves isolated and under great pressure of many kinds: personal danger, psychological stress, social and political isolation and economic hardship. The acknowledgement they receive through workshops, the opportunity to share misery and frustration, hope and determination, the fun and the respite, the oxygen of new energy and ideas from outside, the inspiration which participants can give each other within an encouraging framework, the stories that they hear of the courage and commitment of other people in other contexts – all combine to provide a kind of psychological sustenance and impetus which can be vital.

To give an example, I have worked many times with the group *MOST* (an acronym spelling 'Bridge') at the Centre for Anti-war Action in Belgrade. Together with different colleagues I facilitated several workshops with them when they were just beginning. Our process each time was to spend the first phase of the visit reviewing with the group their current situation and their requirements for our time with them, building together an agenda for the available days – an agenda constantly open to adjustment according to emerging needs. We were able to give them the kind of psychological support described above and, in addition, to provide facilitation for them to become a group, rather than a collection of individuals, developing a common understanding of its purposes and deepening the relationships between each other as its members. We also helped them work through some conflicts and tensions. The sharing of concepts and skills, which could be called 'training', was clearly useful to the group at the outset and provided a basis for the continuing self-development and the external training and peace-making work done by its members. As their work progressed, we provided a framework for evaluation and further planning. Additionally, in the case of this group, our visits provided a link with the world beyond Serbia which meant a great deal then – and still does.

'Training', then, is a somewhat arbitrary word for a collection of functions of which a directly educational element is but one aspect – of greater or lesser importance or predominance according to the particular context. Whereas much of the work I do has an educational purpose, it is clear from this explanation of our work in Belgrade that the provision of emotional space and support can be an important – if sometimes unnamed – function of such workshops, and that the role they play in group formation can be as important as anything else. Even when the participants are unlikely to meet again, they often carry a sense of support and community back into their own situations, a knowledge that there are others who share their aspirations and are working for the same purposes, people they have come close to, and who feel, for a time at least, like invisible companions. For those isolated and oppressed by violent conflict, the sense of being connected to others who support what they are doing fulfils a vital need for solidarity. Armed conflicts tend to erode people's social relationships, networks and communities. Workshops can provide them with new connections and sense of direction, focused on common goals for reducing violence and its consequences.

Training workshops take participants out of the whirlpool of their own daily situation and give them some temporary ground on which to stand and observe it; ground which, if they can make it their own, will become more permanent. Through what is often a very intense time of human exchange, through the sharing of fears and aspirations, the giving and receiving of trust, the weakening of boundaries and stereotypes, participants find support, recognition and encouragement, forging communities which enlarge their own sense of identity and belonging, communities of purpose that transcend the local and sectional interests on which their conflicts are based. Often they forge lasting friendships and networks of cooperation, which are important both psychologically and practically.

CONTENT OF TRAINING WORKSHOPS

Different trainers have different approaches, repertoires and priorities. I hope that mine are not fixed, but constantly developing. I am eclectic in my approach, devising my own ways of using and adapting borrowed concepts and methods, at times inventing new ones. When I use the ideas of other practitioners, I acknowledge my sources. Here I shall set out the range of elements I most regularly use in training. Although some of them (beginnings, endings, evaluation) will always be present, the design of any particular workshop, and the inclusion or exclusion of different elements, will depend on a variety of factors discussed later in this chapter. Methods for addressing different topics are in many cases alluded to within the discussion of the topics themselves (and in some cases further elaborated in appendices to this chapter). Training elements common to many different fields – 'communication skills' for instance – will not be discussed at length, but this is no reflection on their importance. Some general methodological tools and issues will be discussed under a separate heading.

Setting the Scene

The way a workshop begins is important in setting the tone for its continuation and in building trust between facilitators and participants. Where a group is coming together for the first time, the first task, after some welcoming remarks and a brief review of agreed purposes, will be to initiate a process of introduction which draws everyone in and enables each to say the things that s/he most wants others to know about her/him. A well-known method I often use is to invite participants to hold a brief conversation of introduction in pairs, followed by a 'go round' in which each introduces her/his neighbour to the rest of the group. This process has the advantage of immediately involving everyone in a relatively non-threatening way, drawing in the shy and nervous and discouraging the voluble from taking an inordinate amount of time – thereby setting a helpful pattern for the rest of the workshop.

What no introductory process can achieve is for everyone to have told everyone else all they can ever want to have known about them or about their situation and their work. For each to be heard at great length would take an impossible amount of time and concentration. Nonetheless, it often happens that at some later stage in the workshop participants will complain that 'we needed to know more about each other'. (If a workshop is going well, there will never be enough time for all the things that participants want to do. See Chapter 5.) Written information in advance, 'market-places' where participants display information about themselves, evening sessions for them to talk more about their work and special interest groups can all help fill the need, but there is no way that I have yet discovered for everyone to hear everything they want about everybody.

After introductions have been made, I usually ask participants to voice their hopes and fears for the workshop. I prefer 'hopes and fears' to 'expectations', since the phrase invites the disclosure of feelings about the anticipated process and dynamics of the workshop, as well as its content. If, however, I think that 'expect-ations' will be less threatening or more dignified for certain participant groups, I act accordingly. This process provides an oppor-tunity to check purposes again, note adjustments that need to be made to the draft agenda (which I would not usually have printed, but present in a more temporary form on a flip chart). It is also an occasion to offer some comments about what is likely to be possible within the time available.

Since the hopes and fears named often include some about behaviours and dynamics within the group, I move on at this point to a process for developing an agreed list of 'ground rules' or mutual commitments for the behaviour of individuals within the group. Sug-gestions agreed upon (which could all be seen as aspects of respect) typically include: respect for different identities and viewpoints; listening and not speaking while someone else is speaking; not talking for too long and preventing others from speaking; taking responsibility for one's viewpoint or opinion as one's own, rather than claiming to speak for others; keeping to agreements about time, smoking and the like; commitment to confidentiality where it is asked for; shared responsibility for the process. It may seem heavy-handed to propose the making of such agreements, but the process brings into awareness the factors that can make or break group relationships. It draws participants into a sense of shared responsibility for the success of the workshop and provides an important reference point for the facilitators or anyone else who wishes to draw attention to something which is going wrong in the group process. In circumstances where participants are often anxious about what lies ahead, it gives some reassurance. The more tense initial relationships are, the more important this is likely to be.

Although I have on occasion designed the agenda from scratch with participants at the beginning of a workshop, my personal tendency is to have a draft agenda to offer, shaped on the basis of discussions with the workshop's organisers and the 'needs assessment' they have done. That way I feel in some way prepared and the group has a starting place to work from. However, I am ready to abandon it and expect to modify it, in major or minor ways, whether initially or as the workshop progresses.

Use of the 'Stages' Diagram

Figure 2.1, which shows the stages and processes in conflict transformation, summarises my own thinking about different aspects of conflict transformation and provides a framework for my assessment of what workshop content will be appropriate for a given group. I often use it as a reference point not only for my own initial agenda planning but also for the presentation of an outline agenda at the beginning of a workshop and in the explanation of its rationale. It can provide a vehicle for some introductory conceptualisation, presented on a large sheet of paper and remaining on the wall for reference and discussion as the workshop progresses. At other times I do not use it at the outset, but may introduce it in the middle, when relationships between different circumstances and actions need to be clarified. Whenever I use the diagram, I invite participants to relate it to their own experience. I try to make it clear that it is not to be regarded as an attempt to represent the 'truth' about conflict, but as a tool for thinking and talking about it. It engages participants in animated discussion and provides the basis for perceptive observations and new understandings. I shall use it now as the framework for my discussion here of the range of potential workshop content, whose elements I will relate to the stages represented on it. However, I wish to make it clear that the actual content of a specific workshop will depend on the circumstances in which it is held and the composition and needs of the group in question.

Raising Awareness, Mobilisation and Group Formation

In Figure 2.1, the first processes in conflict transformation are named as 'awareness raising' (or 'conscientisation'), 'mobilisation' and 'group formation'. I explain these processes as instrumental in the empowerment of groups who want to act for change, whether or not they see themselves as 'oppressed'. I note also that the workshop in question is an example of an awareness-raising process and that the forms of awareness it raises will include political awareness, awareness of interpersonal and group dynamics, and self-awareness. Early in a workshop I often invite participants to consider their own personal inclinations in relation to conflict, for instance, whether they tend to seek it or avoid it and whether they tend to put relationships before (other) goals or vice versa. (There are likely to be cultural issues to discuss here, as well as personal responses.) I may also invite participants to monitor their own role in the group's dynamics.

I point out that the processes with which the workshop has begun are 'group building' processes: getting to know one another, clarifying common purposes and assumptions, agreeing on group behaviours and responsibilities. Often, depending on the existing experience of participants, I draw from them their own experience of groups and what contributes to or impairs their effectiveness, discussing the different constructive and destructive roles people play in groups and organisations, and different models of leadership and their relative advantages and disadvantages. Frequently I include some work on facilitation, typically using role-play, in which all the participants are asked to observe the different behaviours which are contributing (negatively and positively) to the dynamics of the group, in terms of both 'task' (what they are trying to achieve) and 'maintenance' (looking after feelings and relationships: Jelfs, 1982). Whole workshops or a series of workshops can be devoted to group building and organisational development and strategy, if that is the particular need.

Daily Evaluation

Within any workshop, daily evaluation helps consolidate learning and provides feedback for facilitators. It gives participants an opportunity to reflect on how the group is functioning and on their own individual roles within it, which helps to maintain levels of awareness and responsibility within the group. At the same time it gives them a chance to reflect with others on what they are learning and what they have and have not found helpful. This in turn, fed back to the facilitators, gives them important information on what is working and what problems need to be addressed or what needs to be done differently. In particular, it will help them assess whether the agenda as planned needs changing. Beyond its utility within the workshop itself, the daily pattern of planning, action and reflection/evaluation (repeated on a smaller scale in relation to each agenda item) is part of the learning of the workshop, in that it models one of the principles of effective action.

One means I often use for this daily reflection and evaluation is to set up 'base groups', which provide a kind of temporary home for participants who in some cases know no one else within the wider group. In these base groups a small number of participants, mixed in terms of age, gender, ethnicity, etc., will have the opportunity to get to know each other relatively quickly and can support each other socially in the early stages of the workshop. The base groups are given the opportunity, within agenda time, to meet at the end of each working day and will give feedback on the day's proceedings and on general arrangements, and suggestions for changes or additional activities, either in plenary or, through a spokesperson, directly to the facilitators. Sometimes, however (usually for reasons of time), I ask for evaluative feedback in plenary, for instance, inviting comments under the general categories of 'positive', 'negative' and 'ideas for the future'. While it has the advantages of spontaneity, simplicity and involving (theoretically) the whole group, this plenary process tends to involve some participants more than others, and to leave the facilitator(s) with multiple and con-tradictory demands in an undigested form. The group method is likely to engage more participants and provides more space for discussion. The evaluation it produces is more considered and is communicated in a summarised form which can help facilitators in agenda planning for the following day. It has the added benefit of providing practice in group process, including seeking consensus and acknowledging differences of opinion, and gives time for reflection and conversation between individuals about their own behaviour. Giving and receiving feedback, both affirmative and critical, precisely and constructively, is a useful skill for participants to practise and enables them to help each other learn.

Visions and Values

One way of bringing a group together, which at the same time provides a basis for all the other work on the agenda, is to explore the visions they have for their own societies and global society and the values on which they are based. A good way of presenting visions is to create visual images of them. This can be done by individuals or groups, and the pictures displayed for all to see and explained by those who made them. Men of a certain age, culture or status may find such exercises embarrassing and alternatives may need to be found, but most of the people I have worked with, though some are hesitant at first, soon become absorbed in the ideas they are engaged with. The real danger is that the power of the longings expressed and the gulf between the dreams and the current realities is so wide that the exercise releases a powerful sense of grief and loss. If this happens, it needs to be acknowledged and not brushed aside, and the aspirations expressed turned into constructive and realistic plans for steps forward – however small. (An alternative approach to visions is to start from participants' negative expectations in relation to a particular issue and imagine a radically different unfolding of events and how it could begin.)

One method I use to explore values is an exercise 'borrowed' from John Paul Lederach (1995). It is based on a verse from the Book of Psalms (Psalm 85, verse 10) – 'Truth and mercy have met together, peace and justice have kissed.' The exercise takes the form of a debate between allegorical personae and is designed to explore the tension between Mercy and Truth, Justice and Peace.

Participants choose the value they think most important in conflict, and gather in corresponding groups. They discuss together why this value is of paramount importance to them and choose a representative to speak in the person of that value: 'I am Truth (Mercy, etc.). In the midst of conflict I am of the greatest importance because ... I feel closest to (Justice, etc.) because ... I feel most threatened by (Peace, etc.) because ... .'

Each speaker is questioned by the listeners from the other groups, and members of their own group may speak in their support. The discussion gradually broadens and becomes less formal, as the relationships of support and tension between the different values emerge. Though some participants dislike the word 'mercy', because of its overtones of power and condescension, 'compassion', 'amnesty' or some other variant can be substituted, and the exercise seems to resonate with participants regardless of culture or religious affiliation and never fails to engage. The relationship between the four values can be further explored through 'sculpting' (see below, under Workshop Methods).

Nonviolence and Nonviolent Action

Thinking about violence and nonviolence

Group formation is not only a precondition for successful workshops. It is the first stage in the process of mobilisation for nonviolent action and, again, includes the sharing of visions and values. These are directly relevant for the consideration of nonviolence. Workshop participants do not often have any background in the philosophy of nonviolence or any pre-existing commitment to it, and although it is always clear that the workshops are concerned with nonviolent rather than violent approaches to conflict, rejection of violence may be seen as criticism of a movement for justice. If discussion of nonviolence is part of the agenda, I usually open the topic not by talking about nonviolence but by asking participants about their experiences of violence and in what forms it affects their lives. I often use Johan Galtung's 'violence triangle' (1990) to give shape to the different forms of violence which they describe. I might then open a discussion on possible responses to the different forms of violence that people have experienced, gathering anecdotes and ideas before moving to the notion of nonviolence as such. I might begin this discussion with a descriptive summary of its essentials as I understand them, and invite participants to contribute their own understandings and responses – including objections. This discussion may be major or minor, depending on the nature of the group and the context of the workshop. It will include, undoubtedly, accounts of experience as well as moral and theoretical commentary and will ideally suggest the range and characteristics of nonviolent action.

Discussing violence and nonviolence can be challenging, especially where experiences of oppression have been very bitter and violence is seen as the only way of fighting back. If workshop participants have supported, or been engaged in, violent action, they may be sceptical about the idea of nonviolence, and even hostile to it, seeing it as undermining the validity of violent struggle. Often, however, it emerges that hostility has been a cover for despair. The application of new ideas and models to participants' own situations is often the occasion of real excitement, as they find their experiences taking form and becoming more recognisable – capable of being named, managed and changed. In my experience of working with groups, these really are 'tools for empowerment', enabling participants to make action plans which are both imaginative and realistic. Role-plays can bring their plans to life and help them to see what are the likely obstacles. Although the difficulties and limitations of nonviolent action in a given situation should not be minimised, I have not encountered any group that concluded, after careful consideration, that there was nothing to be done nonviolently, or to reduce or minimise violence.

Tools for analysis and strategy

It will be important at a certain point to move on to test some of the tools of nonviolence in relation to the issues with which the participants are concerned – ways of working towards the visions they have for the future. Sometimes I introduce these tools before any discussion of nonviolence as such, holding the conceptual and philosophical discussion for later, or staying at this pragmatic level.

A tool I have found invaluable for groups wanting to overcome injustice is the set of three analytical models designed by Jean and Hildegard Goss-Mayr (1990) (used also by Fisher *et al.*, 2000. For a full description and diagrams, see Appendix 4.1 (page 123). The example I give is an old one, but the situation it describes is ongoing and it strikes such a chord with so many workshop participants from different parts of the world that I have decided to retain it.) These models provide a framework for 'conscientisation' – that is, for those affected by a situation to understand its nature and reach a common view of what needs to change, using analysis which can lead to effective planning. The first model is used to define the violent or unjust state of affairs (or, less judgementally and more generally, the problem) and to identify the factors or the people who keep it as it is. The second provides a framework for identifying positive goals for new activities, relationships and structures, and the third is used to envisage how a small group of people can attract support and build a movement for change. (In practice, I vary the order in which I present the diagrams.) Sometimes I present them using an illustrative case, like the one used in the appendix; at other times by analysing a local situation or one that has already been discussed in the group.

Action plans

After the analysis of the situation(s) under consideration, they will be asked to consider what action could be taken to mobilise support for change, and increase the likelihood of fruitful negotiations. Although others (most notably Gene Sharp, 1973) have written extensively on the different types of nonviolent action, and some categorisations and examples may help generate discussion and stimulate imagination, I prefer to make very little input at this point. My aim is to encourage participants to recall their own examples (story-telling is very powerful here) and generate their own ideas, and to recognise the endless potential for imaginative and creative action, which may be discouraged rather than stimulated by ready-made categories. Theoretical distinctions between one form of action and another seem relatively unimportant. I do, however, remind participants of the central role of dialogue, and offer some reminders of things to bear in mind when preparing for action.

Confrontation and Conflict Dynamics

Figure 2.1 depicts confrontation or 'open conflict' simply as a pivotal stage which usually has to be passed before conflict can be addressed by the different parties together. In practice, negative dynamics often get the upper hand at this stage of a conflict (as suggested in the more complex version of the stages diagram presented in Appendix 2.1). The best-motivated and best-prepared confrontations can take on ugly shapes and many workshop participants find themselves in the midst of conflicts that have already taken a violent form and negative direction).

Understanding the typical dynamics of escalating conflict and confrontation can help participants to see what needs to be addressed and where it is possible for them to make a difference. In this as in other things, I usually begin from participants' own experiences – what often happens, what makes things worse and what makes them better – and explore the possibilities of averting or reversing the 'together into the abyss' process described by Friedrich Glasl (1997). Sometimes I introduce a theoretical model of conflict dynamics – for instance, Glasl's, or the model developed by Lederach for the Mennonite Conciliation Service (1997) – as a framework to which those experiences can be related. Role-plays offer the opportunity for more immediate experiential learning.

The job of establishing, maintaining and developing constructive communication, which is so crucial to the processes included within the 'resolution' stage of conflict, in practice has to begin here. Within the philosophy of nonviolence, dialogue is fundamental and its pursuit is a constant within nonviolent action at any stage. The consideration and practice of communication skills is discussed below.

Moving from Confrontation to Resolution

A model for problem-solving in conflict

To structure thinking about the 'resolution' stage of conflict, I often use a diagram called 'the iceberg' (though a colleague recently suggested that 'the mountain' might provide a more positive image). I learned it (in action) from a co-facilitator, Tom Leimdorfer, and elaborated it a little. This model can either be introduced in one go or developed stage by stage in the course of a workshop. Either way, I find it a useful framework in which to develop with participants the idea of a co-operative approach to conflict, once parties are ready to begin the search for a way forward – the 'conflict resolution' phase of the 'stages' diagram in Figure 2.1. I would, however, begin by asking participants about their own experience of things that exacerbate conflict or help it to move forward. Respect and communication will usually be at the heart of their thinking, so that the first two layers of the iceberg in the diagram (labelled 'affirmation' and 'communication') mirror what they have experienced as necessary and effective. The third layer, 'co-operation', introduces the idea of problem-solving, outlining its basic elements. (The diagram and accompanying text can be found as Appendix 4.2, page 126.)

Constructive communication and dialogue skills

Having discussed the importance of communication, I often introduce experiential exercises in which participants (usually working in pairs, sometimes with an observer) experience the effects of different qualities of (non-)communication and practise listening and speaking skills, reflecting on what, for them, constitutes effective communication in conflict. Since the appropriateness of different styles of communication is dependent on cultural and other circumstances, I try not to prescribe particular techniques, but rather to sensitise participants to the subtleties and complexities of the ways in which communication happens, and the importance of attentiveness and the will to understand.

While the importance of attentive, respectful listening is readily understood, 'assertiveness' is a concept which is hard to translate, both culturally and linguistically. Participants from most cultures agree upon the need, nonetheless, to make concerns, perceptions and requests known clearly (even if obliquely), without forgetting the likely feelings and perceptions of the listener and speaking with sensitivity and respect.

Obstacles to constructive communication

After using experiential exercises in communication, I often explore with participants some of the things which block it. Among those mentioned will be the effects of cultural differences and language; of strong emotions, such as fear, grief and anger; perceptions of power relations; feelings associated with identity and belonging, such as defensiveness, prejudice and the ready recourse to stereotyping. I often introduce exercises which provide an opportunity for participants to explore the ways in which they deal with these obstacles. In the case of strong emotions, for instance, in order to help them think critically and focus on actual experience rather than some generalised theory, I sometimes use a 'continuum line' exercise (see Workshop Methods, page 110.) For thinking about prejudice, I often ask participants to work in pairs and tell each other both of instances in which they have been victims of prejudice and examples of their own prejudice towards others. The work in pairs is followed by plenary discussion.

Exploring identity

Since many of the conflicts I work with come under the category of 'ethnic' conflicts, and are in part, at least, *about* culture and identity, and since prejudice is related to identity and the sense of self and other, I often explore the two together. I sometimes ask participants to list the different groups and categories of people with whom they identify or to which they belong, then to ask themselves which characteristics of those groups (or categories) make them feel positive and which make them feel uneasy. (This is potentially threatening, especially in workshops comprising different identity groups in conflict. I usually ask participants to work at first in twos or threes, rather than speak out in plenary.) The fruits of this exercise may include the kind of critical examination of culture discussed in Chapter 3.

Reframing

As an aid to constructive communication and thinking in conflict, the concept – and practice – of 'reframing' is of great utility. To reframe something is to find a new way of looking at it, to see it from a different angle, to alter one's view or understanding of it and consequently one's approach or response. When entrenched and negative perceptions have contributed to a conflict and its perpetuation, positive reframing can have a powerful effect. Some forms of reframing I often suggest to participants are:

me versus you → we	coming to see the conflict not as a confrontation between the parties, but as a problem confronting both/all parties which can be (and needs to be) tackled co-operatively (while not losing sight of real differences)
positions → interests (based on needs and fears)	moving from a focus on fixed demands ('This is the only version of the truth about the conflict and these are the only things we can settle for') to a focus on the more specific interests of each side, which may be met in a variety of ways, in the light of the underlying needs and fears that will have to be addressed if a satisfactory and lasting agreement is to be reached
past → present and future	turning attention from the grievances and miseries of the past to current issues and future requirements and options
impossible → possible	discovering possibilities for positive change in situations which have come to seem hopeless and without escape
victims → choosers	shifting the self-understanding of the people in conflict so that they are able to see themselves not simply as the victims of others or of circumstance, but as people who can make choices and be active on their own behalf and with others.

Conflict analysis

Conflicts can be analysed in many different ways and in greater or lesser detail. The purpose of analysis is to achieve a degree of understanding which makes it possible to see what can be done to move things forward. Analysis can focus on the causes of a conflict (what led up to it or the conditions that made it possible for it to escalate into major violence) or on the current situation, or both. It can concern itself with 'hard' issues, like economic deprivation, political oppression or social exclusion, or with 'soft' ones such as people's relationships, emotions and perceptions. It can explore and try to define the power of different parties in relation to each other and to the conflict's outcome: Who are the parties directly involved? Are they relatively homogeneous, or disunited and fragmented? Who else is involved or affected in some way, or perhaps wields some hidden but powerful influence? What are the relationships between the different players – supportive or antagonistic? Who can and should be drawn into talks? How can those who are not included in the talks be taken into account in the thinking and proposals which are developed? Workshop participants can be invited to find their own ways of depicting these things, or they can be offered some method for doing so (see for example, Fisher *et al.*, 2000).

One tool for reframing and analysing a conflict, which I use very often in workshops, is called 'needs and fears mapping'. Although acknowledgement of what has happened (or is perceived to have happened) in the past is often important in the long-term resolution of a conflict, shifting the focus from the past to the present and future is an essential step towards resolving it. Needs and fears mapping is one form of analysis which bridges the gap between emotions and practicalities, and which helps in focusing on the future. I took the idea of this kind of mapping from Cornelius and Faire's *Everyone Can Win* (1989). Though I find the tone of their book too jolly and optimistic for application to some of the horribly complex and violent conflicts I have worked in, this exercise is invaluable for its focus on the fundamental motivations which are recognisable across all boundaries (in line with Burton's human needs theory).

Whereas the first of the Goss-Mayr analytical diagrams provides a tool for one group to reach its own understanding of a problem or injustice and who or what supports it, needs and fears mapping provides a framework and process for the representation of the viewpoints of all parties to the conflict in question. Although I usually introduce it in relation to the resolution stage of a conflict, it is also useful at the mobilisation stage. It can help a group that wishes to promote its own cause to weigh the justice of its aspirations in relation to the needs of others. It can also help the group to understand those it wishes to confront. To begin to think about another party in terms of their needs and fears is a step towards empathy – and away from demonisation. (I remember one very militant workshop participant who told me ruefully that doing this exercise had made him see the conflict in question in a new way.)

Needs and fears mapping does not attempt to define the relationships of the different parties (or 'stakeholders') either to each other or to the problem, nor does it address the question of power, but within its scope and for all its simplicity, it can be extremely illuminating in revealing the complexity of the conflict. For a full explanation of this exercise, see Appendix 4.3, page 128.

Negotiation and Mediation

If there is to be constructive dialogue in conflict, the primary need is for articulate and constructive spokespeople for the different parties. The work on communication will do something to address this need. The application of conflict analysis and of problem-solving approaches to negotiation through case studies, role-play and discussion will help to develop ideas and skills for effective negotiation for inclusive solutions. (For another approach to negotiation – 'principled negotiation' – which overlaps with the 'problem-solving' approach, see the work of Roger Fisher and William Ury, 1981 and elsewhere.)

I often use the 'needs and fears maps' (described above) that have been developed around participants' own situations as the basis for negotiation role-plays in which participants can test and develop their communication skills and their capacity for both analytical and imaginative thinking.

The exercise of looking at the needs and fears of different parties to a conflict is also an excellent way towards understanding something of the role of a facilitative mediator, whose job will be to help manage the strong emotions surrounding the conflict, remaining sensitive to the needs and fears of the different parties, as well as to facilitate the problem-solving process as such. Sometimes I suggest beginning the role-play without a mediator, then continuing or doing it again with one, to see what difference it makes.

Although relatively few workshop participants are likely to become mediators in any formal sense, most will play some kind of informal mediatory role, either between individuals or within and between groups, so I often give some time to the topic of mediation, even in introductory workshops. Serious training for negotiation and mediation takes a good deal of time – far more than is available in most workshops. But even an introductory exploration of approaches, roles and skills can be useful for an understanding of the different possibilities for moving a conflict further, and can sharpen the general skills of participants for behaving constructively in situations of tension and complexity. Mediation role-plays provide an excellent opportunity for developing an understanding of the skills and sensitivities that are essential to most, if not all, forms of action for conflict transformation. They also help in the development of general facilitation skills. Mediator functions to which I draw attention include:

- being a focus for and generator of trust, offering respect, confidentiality and understanding, and helping the parties believe that constructive dialogue is possible and that a way forward can be found
- helping to reframe the conflict as a common problem
- helping to create an atmosphere in which anger and disagreement can be expressed but also managed, exercising when necessary the authority which has been given to the mediator by the parties to maintain an agreed process and direction
- helping to clarify, connect and conceptualise issues and options, encouraging the parties to be clear about what they need and what they can offer, and helping shift attention from the past to the present and future
- encouraging imagination and analysis in relation to the generation and evaluation of options and their implementation.

A list of the qualities needed by a mediator in order to fulfil all these facilitative functions would indeed be long, but self-awareness, sensitivity, patience and a steady nerve would have to be included, along with the ability and will to listen and understand, keeping personal opinions and preferences out of the process, being strictly non-partisan and guiding the process but not the outcome of the negotiations.

As I suggested in Chapter 3, the relevance to different cultures of facilitative mediation models emanating from the West is much debated (see, for example, Lederach, 1995). They run counter to most traditions. Their usefulness in participants' own contexts needs to be discussed, along with specific questions about the advantages and disadvantages of mediation, as against arbitration; about who would be acceptable as a mediator; about the appropriate context for a face-to-face meeting; and about the social forms that would characterise the meeting itself.

Whether formally or informally, the following stages would usually need to be included in a face-to-face mediation meeting of this type – though they would not necessarily follow each other in an orderly manner and could take many meetings to complete:

- introductions and agreement about agenda and process: 'What we are here for and how we will proceed'
- each party gives its own account of the conflict: 'What has happened and who has done what and why'
- exchange of views and positions – and expressing the feelings behind them: 'What I think and feel about what you said, and about the current situation and the future'
- clarifying the parties' needs, specifying issues, and agreeing which are the most important to either side
- generating and examining alternatives for addressing the problems
- reaching agreement (and if appropriate writing it down), including any monitoring/review process for its implementation
- if appropriate, and according to culture, some social exchange, symbolic act or ceremony (toast, handshake, exchange of gifts, etc.).

Since facilitating face-to-face meetings is often a relatively small part of a mediator's work, I also explore with groups the question of how parties can be helped to reach the point where mediation seems possible, practising preparatory meetings and shuttle mediation. In most real-life violent conflicts, preparing the ground for face-to-face negotiation or mediation is the most difficult and frustrating part of the conflict transformation process. It is closely related to the development of peace constituencies.

Peace Constituencies

In any consideration of negotiation and mediation at the socio-political level, the relationship between leaders and their constituencies is another important area for consideration. The preparation of constituencies for a move towards negotiation and settlement, and ways of including them in the process, is vital in deep-rooted and complex conflicts. In a general workshop these issues can only be touched on, but in workshops whose participants are at this stage of a conflict, they may be more important than anything else. (The Goss-Mayrs' analytical models can be useful here, too, for instance, if the problem focused on is resistance to serious dialogue, or the lack of the trust necessary for settlement.)

Processing the Past; Reconciliation

Reconciliation, which appears as the final phase of conflict resolution in Figure 2.1, is likely, after violent conflict, to be a slow and complex process. (If reconciliation means the restoration of workable, decent relationships, 'conciliation' would often be a better word, since such relationships may not have existed within living memory.) At the societal level it involves the reweaving of the collective fabric of interaction. In situations where a settlement is imposed (as is so often the case), it is likely to be hard to implement, containing new injustices which need to be rectified. Positive relationships at the interpersonal and inter-group level depend on satisfactory structural relationships. Working as I do with people who have suffered greatly in violence and turmoil – who in some cases have lost most of what they ever counted on – I realise that for some the notion of full recovery is meaningless. At the same time I know many who seem to have performed miracles, and who have re-created their own sense of meaning through helping others to recover and to rebuild society. The human will to make a life, even out of ashes, seems to be both deep-seated and resilient.

I often open the topic of recovery and reconciliation by asking participants to reflect on their own experiences of having been hurt in different ways, large and small, and how they have been enabled to recover and move on – or, if that has not been possible, what has prevented them. These personal reflections (often painful) provide the basis for a discussion of larger political processes and needs, the ways in which experience at the personal level can and cannot be applied, and the additional factors which have to be addressed. Case studies, for instance, from Latin America, South Africa and the Balkans, provide further food for thought, revealing the importance of physical security; of support for those who have been traumatised; of acknowledgement of what has happened and of reparation; of the problems associated with the return and reintegration of refugees. They demonstrate that small-scale mediation between former neighbours may make impossible encounters possible and that human rights monitoring and advocacy will be important, as well as work to create a public climate for reconciliation and co-operation. They show the role the media can play and the importance of education. Participants may well have their own examples of many of these things, and of the contribution of community relations work and different bridge-building initiatives.

Long-term Peace-building and Peace Maintenance

Most of the elements of recovery listed above constitute tasks for the long term. They need to go hand in hand with economic recovery and the rebuilding of physical, social and political structures and infrastructures. The establishment of a culture and mechanisms conducive to constructive conflict management will also be vital, as part of the more general establishment of democratic structures and processes. In a training workshop, these needs can be reviewed and discussed and issues of particular relevance to participants explored further.

The subject of democracy is a vast one to enter. Nonetheless, useful work can be done in identifying, challenging and developing key concepts about the nature of democracy and its different forms, and related skills, actual and potential, at different levels and in different contexts; different styles of leadership and their relative advantages and disadvantages; the relationship between public, private and voluntary sectors; the role of the media and other forms of communication; the role of law and police and other mechanisms for handling conflict.

One concept which I find invaluable in workshops about democracy, political thinking and participation is Isaiah Berlin's notion of 'competing goods' (1998). Just as, in Lederach's exercise described above, justice may be in competition with peace, or mercy with truth, so, for instance, social security may be in competition with personal freedom, and the balance which is struck between the two will determine policy on law and civil liberties, and on levels of taxation for the provision of public services. Other values to be weighed against each other include those of opposition and consensus, of strong central government and the devolution of powers, and the relative needs of majorities and minorities.

When it comes to the consideration of what is required for sustainable peace, conflict transformation takes its place in a wider picture and the boundaries dissolve, merging with questions of governance, law, constitutional reform, policing, social security, economy and the like. 'Civil society development', which is seen as essential to democracy, could also be regarded as a field in its own right, and training in management and organisational development is an important aspect of 'capacity-building'. However, it should not become another prescriptive element in a Western blueprint for other societies. The business of participation in public affairs may be done in a wide variety of ways, and it is those who live in a particular context and culture who will know what those ways are. However, the processes of thinking about groups and their effectiveness outlined above, since they are elicitive in character, are relevant to any work on collective action, as are many of the other elements of workshop content.

Choice of Role in Conflict Transformation

If they are to maximise their power to influence the course of conflict, workshop participants will need to identify the most appropriate roles for themselves as agents of transformation. Such is the strength of emphasis which has been given to third-party roles in the field of conflict resolution/transformation that workshop participants sometimes see mediation as synonymous with conflict resolution/transformation. This is confusing and counterproductive. I therefore discuss with participants in workshops the range of constructive roles that can be played at different stages of a conflict. When they are considering events in their own town, country or region – a conflict in which they are willingly or unwillingly implicated – they may need to recognise that they are effectively disqualified as mediators by their own ethnic (or other) identity or perceived allegiance (though it is not impossible for insiders to be trusted to act impartially). There will, however, be many other possibilities for action: partisan roles, acting for one side; bridge-building roles, forging links between one side and the other (promoting dialogue, for instance, or running bipartisan projects or developing bridge-building structures, such as inter-religious committees or inter-ethnic councils); and non-partisan (or omni-partisan) roles, in education and training, in advocacy for human rights or for the mobilisation of a peace constituency to move those in power from the pursuit of victory through violence to a commitment to settlement by negotiation.

When the participants are all from one group which has an existing policy and programme, their need may be to become clearer about their aims, and to examine any related difficulties and dilemmas, or proposed changes of activity. If the group is still at the formation stage, or is a composite group, an exploration of possible roles to take may be one important function of the workshop, enabling participants to consider their potential contribution, in the light of the nature and stage of the conflict with which they are concerned, asking themselves which aspect of the overall situation they wish to try and address, at what level and in what way. I suggest that they take into account not only the needs of the situation but their relationship to it – their motivation, convictions and values, and the skills, capacities, resources and influence they can bring. These reflections can take place through individual thinking, discussion in pairs or small groups and plenary discussion.

Ending a Workshop

Final evaluation

Whereas daily evaluation is important for participants' ongoing learning and facilitators' guidance, a final evaluation process helps to summarise the learning and other benefits that have been achieved during the workshop as a whole, as well as any perceived deficiencies. This may provide useful information for the facilitators and organisers. Final evaluations may get to the essence of what has been valued in a way that the more particular daily feedback has not. If base groups have been used, a final group meeting prior to the closing plenary session will help to prepare their input for this overall evaluation and give members a chance to take leave of each other. In theory, additional written evaluations may give an opportunity for individual participants to reflect at leisure, free from social pressure. In practice, they are often either written in haste and given in before participants leave, or never sent in, but in cases where a substantial number of the participant group has been convinced of their importance and takes them seriously, they can offer useful clarifications and additional insights. (I will return to the more general question of evaluation in Chapter 9.)

Motivation, commitments and 're-entry'

Just as the motivations for conflict, and the responses it generates, are not only rational, neither are the resources needed to transform it. There is something deeper: the source of the will to find and make meaning in the midst of apparent senselessness; the longing and determination to act compassionately in the face of harshness and cruelty; the courage to keep going when it would be easier to give up; the values that persist in spite of everything. I can only name this source as 'spirit'. (To refer to it as 'emotion' will not do, because although emotion is involved, that word suggests something transitory and relatively superficial, whereas the energies and dispositions to which I am referring are deeper and more long-lasting.) To nurture this dimension of participants' being and experience (and so maximise the workshop's potential) is a difficult and important matter. Participants draw inspiration and encouragement from each other and from facilitators, and from the ethos created within the group. It can be augmented by play, by music, by drawing and painting, and by the quality and depth of the things that are said. In a religious group it can be expressed in meditation or worship. It can also be nurtured by the setting of the chosen venue and by small signs of care in the arrangements that have been made.

The idea of giving participants a spiritual boost is not often (if ever) mentioned by workshop organisers – many of whom might be embarrassed by the suggestion that such an intention formed any part of their purpose. But I believe it is often a significant part of a workshop's outcome, and that even organisers who would not speak of such an intention do in fact hold it. If the situation of participants is dispiriting and they have felt downcast and defeated, the workshop experience can be like water in the desert.

At the same time, it is vital that any new hope that is generated should be based on real resources and possibilities, not self-deluding. The further one gets from reality, the easier it may be to create a sense of temporary euphoria. Workshop organisers and facilitators should not be purveyors of easy recipes and delusions. Since the world of workshops is both intense and contained, it is important for participants to have an opportunity to think what their new ideas and enthusiasm – sometimes quite euphoric – will mean in the cold light of day, when they return to the realities of life at home and work. It may also be important to discuss likely problems of isolation, difficulties of explaining new ideas to family and friends, discouragement, and exhaustion or 'burnout'. (These issues may in some circumstances form a major part of a workshop.) But although new ideas and intentions formed by participants should be carefully examined as the workshop draws to a close, and tested in the light of the reality to which participants will return, hope remains essential to the changing of reality. It is good to explore ways of sustaining it.

A final round for voicing briefly what the workshop has meant to each participant and how s/he intends act on new understanding, ideas and plans can be a powerful and satisfactory way of ending a workshop.

DESIGN OF TRAINING WORKSHOPS

Selection and Arrangement of Elements

Workshops vary in length, usually more according to what is possible than what is ideal. It is often hard to decide how best to use the time available – how much to try and include and how many working hours to schedule in a day. It is important to leave enough informal time. (I have established a preferred pattern of five-day units, with a free afternoon in the middle and four one-and-a-half-hour sessions in each day, but such an amount of time may be impossible for participants.) The venue will also have an effect on what can be done and how. For instance, temperature and other matters relating to physical comfort need to be borne in mind when the length of sessions is being decided, and the number of small groups that participants can be divided into may be determined to some extent by the number of rooms available. Unless they are part of a regular course, which is offered publicly and chosen as potentially useful by self-selecting participants or their employers, workshops are ideally tailor-made for particular groups and circumstances.

If the participant group in question has very particular needs and interests, the workshop's focus is likely to omit much of the conflict transformation menu I have outlined, and to concentrate on a more thorough elaboration of one particular aspect of it. Similarly, if the workshop is one of a series, or forms part of a longer course, or caters for people who have already received substantial training and wish to specialise, the agenda will have a narrower focus. Although an introductory workshop may include something from most of the content elements outlined above, finding the right ordering for the agenda of a particular workshop can be important. If a group's most urgent preoccupation is with addressing injustice, it makes sense to begin with content elements related to group formation and mobilisation, violence and nonviolence, moving later to the elements associated with 'conflict resolution' and finally to recovery and healing. If their first interest is in dialogue and resolution processes, I make those my starting point. If participants are struggling with the after-effects of a conflict which has already reached some kind of settlement, recovery and peace-building should provide the main focus of the workshop, and its starting point, with campaigning or organising skills, if they are needed, coming in later, in the service of those aims. In a geographically mixed group, there may be no such common context, and the choice of where to begin is therefore somewhat arbitrary. In such cases, I try to sense whether the overall emphasis required is on 'conflict resolution' or 'nonviolent action' and choose my starting point accordingly.

If there are different groupings among the participants, divided along ethnic lines, it could be important to take questions of identity and stereotypes relatively early on, and to consider which topics and processes will help build trust in the group. On the other hand, working on differences of identity may be too threatening to begin with and may therefore need to be kept for later. In other words, the choice and arrangement of different elements in such a workshop should be ordered and proportioned according to the estimated or known needs of a particular group, and modified as the workshop progresses.

Attention to Group Energy

For the workshop to be effective, it will be necessary in its design to pay attention to 'maintenance' as well as 'task', devising a well-balanced agenda in terms of work and relaxation, and using a varied methodology (groups of different sizes, discussions, visual representations, role-plays, etc.), being aware of the different kinds of energy and attention required by each. Games (as against learning exercises) can also play a part. Although I do not use them with all groups, they can contribute to the pleasure, informality and warmth of the workshop, at the same time as helping to keep participants awake and engaged (especially in the 'graveyard' session after lunch!). While their use requires sensitivity, bearing in mind the personal styles and culture of participants, they can help release tension, break down barriers and renew energy. Sometimes, perhaps surprisingly, it is in the most tense atmospheres and with the most traumatised participants that they can be most valuable.

WORKSHOP METHODS

In my discussion of workshop content, above, I have often indicated how I might explore a given subject with a group, and in some cases referred the reader to models set out in appendices to this chapter. My choice of methods is, no doubt, idiosyncratic, though not without logic. My repertoire has been built up over years of practice and continues to grow and change. Most of the methods I use will be familiar to many readers: plenary 'rounds', when each participant has the chance to speak without interruption; plenary discussions; story-telling; short presentations in plenary, following, or followed by, discussion of experience, often done first in small groups; listening exercises or exchange of personal perceptions and experiences done in pairs; 'brainstorms' on a given topic, listing, without comment, participants' ideas about some question on a flipchart, followed by discussion; giving participants some task to do in small groups, with specific questions or 'models' as a framework, to be presented later in plenary. Because these methods are widely used and relatively straightforward, I will not describe them here. Relevant manuals on workshop methods are given on page 122 for readers requiring more information. I want, here, to discuss 'experiential' exercises such as role-plays, simulations and 'sculpting' which, though frequently used, are also quite difficult to facilitate successfully. I want also to introduce a simple but powerful method for stimulating self-examination and introducing subtlety into the discussion of issues which tend to evoke simplistic or polarised responses – an exercise which is, perhaps, less well-known than the others mentioned above and with which I shall begin.

Continuum Lines

This method for stimulating awareness and discussion is called variously 'continuum lines', 'spectrum lines', 'barometers' or 'line-ups'. I like it because it involves all participants at once, at least physically and mentally, if not verbally. I also like it because it allows for and even imposes a level of subtlety in addressing and weighing issues, so that absolutes become the exception rather than the rule. It involves asking participants to take up a position, literally, in relation to some question, somewhere along an imaginary line indicated on the floor, usually between one wall and the other. Each end of the line indicates one extreme in the polarity. The question might be about competing values, for instance, 'If you have to choose, which do you value more highly, peace or justice (or freedom or security)?' Those who strongly and unequivocally favoured justice in all circumstances would stand at one end of the line and those who had similarly strong feelings for peace at the other. Most would stand uncertainly at different points along the line in between and most would be wanting to say something by way of explanation: 'I've chosen to stand here because ...'; 'It depends what you mean by ...'; 'It depends on the circumstances ...'; 'It's a stupid question because ...'. Some may choose to represent their position by walking up and down. The discussion is always animated and usually continues long after the participants have crept back to their seats or sat on the floor. I use this exercise also for exploring prejudices, outlining a situation involving someone from a group against whom prejudice is possible or likely and asking participants to take up a position somewhere between the poles of claiming complete freedom from prejudice and the admission of horrible guilt! The scenarios can be given in quick succession, with all discussion coming after, or a few participants at different points along the imaginary line can be asked to speak each time; or again, participants can be asked to meet in small groups to discuss the reasons and feelings they had and their observations about the exercise and what they learned from it about themselves and the issue in question. I also use continuum lines as a tool for exploring the notion of 'competing goods' in democracy (see above).

Since 'continuum lines' help participants to evaluate their own attitudes and behaviour they can also be used, for instance, to reflect on how they deal with anger and fear, or whether they tend to be passive or aggressive in conflict, or whether they put relationships before other objectives or vice versa. In short, they offer a lively way of engaging in challenging reflection about their own attitudes and patterns of behaviour, and about their values and judgements. They can also be used as a tool for making difficult group decisions.

Role-plays and Simulations

Role-plays are not games or performances. They are processes based on invented scenarios and roles for testing out different behaviours and their effects, both internal and external. They offer the opportunity for taking risks, trying new approaches and making mistakes in a way which would be impossible in 'real' situations. I put 'real' in inverted commas because role-plays can be very real. That is up to the participants, who are sometimes sceptical, or reluctant, or both. The more they are able to immerse themselves in their given role, the more they and others will learn. At the same time, if they fall too much in love with their role and cannot keep a simultaneous watching eye on themselves, they may overplay it and obstruct the group's learning. If role-play is something new for participants, all this needs to be explained to them.

The roles distributed or chosen will be of two sorts. One is to be oneself within the given scenario, doing one's best to act constructively in an imagined role, for instance, as a mediator or trouble shooter, or as a passer-by intervening to defend someone from attack, or as a campaigner delivering a petition. The other kind of role is just that: playing someone other than oneself – different, possibly hostile; behaving realistically from within an imagined personality as well as an imagined role. A great deal can be learned from the experience of playing either kind of role, but they are very different. This, too, needs to be understood by participants. It is the second kind of role that carries the danger of overplaying. This can have particularly detrimental effects in 'role-reversal' exercises, where the purpose is for the players to put themselves in the skins of the 'other'. If this task is not thoroughly understood, it can instead produce caricatures and the reinforcement of stereotypes. It is common for one or both of the adversaries in mediation role-plays, for instance, to be far more hysterical or intransigent than they would be in real life, rendering the mediator's task impossible and preventing him or her from learning anything about what is likely to 'work' or not. On the other hand, if the player is unable or unwilling to give their allotted or chosen character any life, or enter into the feelings of the imagined person, he or she will contribute little and learn less.

Many participants are nervous of role-plays and I do not force anyone to take part in them (or in anything else for that matter!), though I do emphasise that role-plays are not proficiency tests but opportunities to learn about what works and what does not. It can be useful to have observers who give feedback to the players when the role-play has been concluded. They may notice things that have escaped those who were involved. I have on occasion invited observers to pass notes to players, suggesting different behaviours or proposals for them to use or ignore. Occasionally I 'freeze' a role-play so that its dynamics can be discussed and new approaches tried after discussion. And I sometimes call a break for players to 'caucus' with those they represent in the role-play.

It is important to allow enough time to discuss the role-play or the learning will be lost. There is also a danger that negative feelings may be carried over into the rest of the workshop, without a process of detachment from the role-play. I usually ask participants in the role-play to stay in role for a moment, inviting each to describe briefly how they were feeling at that moment when the role-play was stopped. Then I invite them to leave their role behind – maybe change seats as a sign that they have done so – and reflect on what they noticed in the course of the role-play. When they have all spoken I ask observers for their comments and finally add my own. All this means allowing at least as much time for debriefing as is spent on the role-play itself. The preparation of the role-play will take a similar amount of time: firstly devising or describing a scenario which is well-suited for the desired learning (this is best done with some or all of the group to ensure that it is realistic for them in all respects); secondly allocating roles; and thirdly giving time for the players to think themselves into their role.

Twenty minutes is quite a long time for the role-play itself. A mini role-play, testing out what happens in a very brief interchange, could last only two minutes. Again, I explain in advance that the purpose of the role-play is not to complete a task but to test some process, so that to be stopped when the matter in hand is incomplete is not a sign either of failure on the part of the players or impatience on the part of the facilitator, but a necessary measure if the process of learning is to be completed.

Whereas role-plays are focused on very specific and limited scenarios or episodes, simulation exercises are designed to explore the wider dynamics of a complex process or system. (For example, a simulation exercise could be used to see how different aid organisa-tions might interact with local warlords, with each other and with UN representatives, at a particular moment in a 'complex emergency'.) The focus of the experiment and of what is to be learned from it is on functional rather than on interpersonal dynamics, though it may take in both – and the interaction between the two. A great deal of time is needed for preparation, for the exercise itself and for the debriefing – for instance, one or two days.

Role-plays and simulations are demanding for facilitators, in terms of organisation, attention and sensitivity. They are also demanding for participants. They are, however, when they go well, excellent vehicles for learning, being real enough to uncover aspects of interaction too subtle for theory but not so real that participants are unable to take enough distance to reflect on what has happened and therefore to learn (see, by contrast, the discussion of 'experi-ential learning' at the end of Chapter 6). As with other aspects of training, the best way of learning how to facilitate these processes is through experience as a participant and by working with an experienced colleague.

Sculpting

One powerful way of exploring relationships between individuals and groups is to invite them to 'sculpt' them. One participant or a group of participants is invited to use others as their sculpting material, arranging them in different positions in relation to each other. For instance, in a workshop described in Chapter 8, different 'sculptures' were made to depict the relationship between Serbia and Kosovo/a, and in another to depict regional relationships more generally. The method may also be used to explore the relationship between different values, as in the Lederach exercise described above, or between individuals or categories of people within an organisation. It may be that a person representing one group is placed in an upright position, with her or his foot on the crouching form of another, or be in an attitude of menace while another cowers; or one representative figure may be separated from the rest of the sculpture. While the figures are still in position, or immediately after, the participants involved are asked what it feels like to be so placed, and invited to experiment with changing positions. Those who are observing are asked for their perceptions and suggestions and may be invited to change elements in the sculpture. The exercise may be repeated, with different participants or groups of participants acting as sculptors and sculpted.

This exercise can evoke strong emotional responses, so needs to be used with care. It can, however, unlock feelings and insights which otherwise remain inaccessible, and bring participants together at a very human level.

FACILITATION OF TRAINING WORKSHOPS

Facilitation Tasks

The approach to training which I am describing is one in which the role of the workshop facilitator or 'trainer' is more elicitive than didactic. It is her/his job to devise and facilitate mutually educative processes, drawing out and giving new shape and meaning to the experience-based knowledge that is already present within the group. Facilitators and participants work in partnership. For the facilitator this involves:

- taking lead responsibility for the design of the workshop's content and the methods used for exploring it

- planning the daily agenda in response to the emerging needs and dynamics of the group, negotiating changes with participants
- explaining the workshop's content and agenda, and particular exercises/processes within it
- facilitating those processes and ensuring that the agreed content is covered within the time available
- eliciting, clarifying, collating and summarising ideas and experiences, and offering concepts to encapsulate them ('clarify, connect, conceptualise')
- presenting ideas and information, and recounting the experiences of others (occasionally one's own)
- on the basis of agreements made with and between participants, helping to manage power relations within the group and ensuring that all have a voice
- paying attention to collective and individual feelings and energies within the group, responding to needs for stimulus, breaks, emotional support, spiritual uplift, fun, food
- as appropriate for learning and for the effectiveness of the group process, challenging things that are said and naming and facilitating the management and transformation of conflict within the group
- presenting ideas
- assisting the process of ongoing reflection, feedback and evaluation.

Co-facilitation

It is highly desirable to have more than one facilitator in any kind of workshop. My strong preference and usual practice is to share the work equally with a colleague, planning the agenda with him/her and facilitating each session together. When there are two facilitators, they can take it in turns to lead while the other takes more of a back seat and is freed to observe in a broader way what is going on in the group, in terms of both task and maintenance. Sharing the facilitator role and power in this way not only allows for the pooling of skills and different areas of expertise and experience, it also models power-sharing and helps to create an ethos of co-operation. If the workshop extends over several days, the chance to take a back seat at times, and the support of knowing that one does not carry all the responsibility, can make a great difference. If a conflict flares within the group, it may be important to have someone spend time with one or two participants while the other continues with the rest of the group. In any case, moral support is very welcome and a chance to debrief together can help in adjusting perspectives and reaching decisions about how to proceed. In general, two heads – and two voices – are better than one.

Training for Trainers/Facilitators

Although this can be seen as the most effective means of skills-transfer and capacity-building (Fisher, 1997), it is also a complex and lengthy process. It involves the development not only of a thorough understanding of concepts and skills for conflict trans-formation, and of the role played by training and facilitation, but also of the skills necessary to carry out all the tasks enumerated above: skills of self-awareness, attentiveness, sensitivity, clear and diplomatic communication, flexibility, firmness, good humour; the capacity to think and act quickly, to collate and summarise ideas 'on the hoof', to present complex ideas simply but not simplis-tically, etc.

To cover so much ground and support the development of so many skills requires either an extended training course or a multi-plicity of workshops. Often those who become trainers have participated in many workshops, maybe including some training for trainers, and begin their own practice by working with a more ex-perienced trainer, or in a 'safe' environment – for instance, in a short workshop with friends or colleagues as participants.

AIMS OF DIALOGUE WORKSHOPS

Dialogue workshops bring together participants from different 'sides' in a conflict, and their primary focus is on the relationship between them and the conflict which divides them. The range of aims for dialogue workshops is quite wide, including breaking down stereo-types, broadening perspectives, developing common understandings and aspirations, and identifying possibilities for parallel or joint action. Sometimes the dialogue is focused on rather general issues, such as the development of democracy or the role of religion in inter-ethnic relations. A shared interest is used as the basis for a kind of oblique encounter, just as it is when 'training' is used as a way in to dialogue or as a vehicle for it. Sometimes the focus is on the conflict itself, and can lead to action within their own sphere by par-ticipants or (in the case of problem-solving workshops) influence by them on others who have the power to act. The support and inspiration identified above as an important element in training can also be significant for participants in dialogue workshops.

CONTENT, DESIGN AND FACILITATION OF DIALOGUE WORKSHOPS

The 'scene-setting' processes outlined above in relation to training workshops (welcoming and introductions, discussion of expectations, agreement on ground rules) outlined for training workshops are of equal (arguably even greater) importance in workshops for dialogue. In addition, I find it important to acknowledge the courage and trust which participants have demonstrated by their very willingness to meet each other, and to do everything possible to address fears and insecurities, so that they feel free to enter fully into the process. The higher up the social or political ladder they are, the greater the importance of meeting the demands of protocol and symbolic (as well as real) even-handedness will be. Agreements about confidentiality, always important, will also take on greater significance.

Most of the process will take the form of facilitated discussion, sometimes with the parties working separately, sometimes together. Some of the tools used in training can offer a helpful framework in this kind of context too. In dialogue workshops, skills and methods are used not as exercises but 'for real', applied to the external conflict represented within the group by its participants. For instance, in a problem-solving workshop, participants may be invited to set out their positions, then be given an explanation of the process of problem-solving and asked to engage in it. 'Needs and fears' mapping has proved invaluable in my experience, with each group presenting its own definition of the conflict and an analysis of the needs and fears of its members. At a later stage in the workshop joint visions may be developed, and the Goss-Mayr models for analysis and strategy may be used to explore how the parties can address obstacles to problem-solving on their own side or in the overall situation and used as a tool for action planning.

Theory about conflict and conflict transformation may be introduced into dialogue workshops as and when it is felt to be appropriate. It can help participants to take some distance from their involvement and see the conflict from a different perspective. Sometimes people from different regions of conflict may be invited to take part and to make some input about their experiences and what they have learnt from them, which can help participants to see their own conflicts in new ways. The use of exercises like continuum lines and role-plays can work well in more informal dialogue workshops. Since they require a substantial degree of personal exposure, it is important to judge their 'safety' or suitability in a given context. In some circumstances participants find it undesirable or unsafe to speak and act openly as an individual and want to maintain a united front with the others from their group. Given enough time, and success in creating a safe atmosphere, this need for control may recede, so that greater openness to new ideas is possible.

In view of the need for trust to be established, the order of the agenda in dialogue workshops is particularly important and involves difficult decisions, in particular, whether to plunge straight in, by beginning with the central or most fundamental issues, or to enter the water more gradually by starting with less threatening matters. This is a question for fine judgement in each particular context. It is also, for the facilitators, a matter of personal style and feel, depending on what they think they can handle themselves, and enable participants to handle constructively. They may choose to consult with participants or to rely on their own professional judgement. It is their job to create a safe environment, psychologically speaking. (One option, as mentioned earlier, is to hold a training workshop with the group first; then, on the basis of the trust established, to go to the heart of the matter in a problem-solving process.) In circumstances where participants have already expressed their readiness to address a conflict directly, I prefer to spend time on a thorough introductory process, but then move as quickly as possible to the issues that stand between the participants, so that the maximum amount of available time can be used for the real business of addressing those issues. If they are postponed or avoided for too long, time is wasted and the dialogue stays at a surface level where trust is apparent but not real, since everyone is aware of the avoidance. However, a series of workshops or other initiatives, over a long period of time, is often needed before such a point can be reached.

Sometimes participants want to tackle the most difficult and important questions, but at the same time are afraid to do so and persist in avoiding them. In this case, if no participant does so, it may be helpful for facilitators to name this avoidance (tactfully) and suggest a process for overcoming it.

It will be clear from the above that flexibility about the agenda and the workshop's flow is particularly necessary in dialogue workshops. It will also be clear that the task of facilitators in dialogue workshops (which is a mediatory one) is delicate and demanding. It also holds the danger of degenerating into manipulation. I believe that transparency – keeping participants informed of what the facilitators are thinking and the reasons for what they are proposing – is both the most respectful and the most effective way of dealing with this danger.

The dynamics and changing needs of dialogue processes will be further discussed and illustrated in Chapter 8.

SOME PRACTICAL MATTERS

Some practical questions are important not only for practical reasons, but because they send messages, intentional and unintentional, which have a serious impact. One such question is that of group composition, another is that of language, and a third is that of location.

Group Composition and Numbers

When invitations to a workshop are made, it is desirable that the group, if it is not a pre-existing one, should be composed with care, with attention to balance between different categories within the given constituency – such as men and women if it is a mixed group, participants from different parts of the world if it is an international workshop, or those identifying themselves with different ethnic or religious groups if the workshop is regional or local.

The ideal size of a group will be influenced by circumstance (cost and demand, for instance, and available venue) but my usual preference would be for 15–20 participants, and I would not normally choose to work with a group of more than 25. (Since it often happens that a few participants fail to materialise, if this is likely a few more may be invited to allow for this.) It should in any case be taken into account that the larger the group, the longer the time needed for all participatory processes.

Language

In dialogue workshops between different ethnic or cultural groups, where language is often part of what is at issue, the symbolic importance of language is strong, and the use of a third language or interpreters is necessary. In international training workshops, seminars and conferences, either some people work in a second or third language, or there is a need for interpretation. Sometimes, even with interpretation, many participants have to work in a language other than their 'mother tongue', which reduces their ability to communicate and can be frustrating and disempowering. The very expression 'mother tongue' (often used by participants) conveys something of the security and insecurity associated with language, and its symbolic nature in relationship to identity and culture. Moreover, questions of language are often questions of power. It is not by accident that the working language of so many seminars, conferences and workshops is one of the colonial languages, with English in the dominant position among them. Even without the weight of colonial history or the tensions of inter-ethnic conflict, competing language needs within a group raise challenging questions about the relative rights of minorities and majorities. Although there are no simple or complete answers to these issues, it helps to recognise their sensitivity and their importance for full participation. A combination of recognition and practical care, on the part of facilitators and participants alike, will usually make it possible for the difficulties to be managed, providing excellent opportunities for experiential learning. (For instance, in one group I worked with, the partial relinquishing of power by linguistically powerful English-speakers who, despite the weakness of their French, volunteered to work in French-speaking groups, was greatly appreciated and had a profound impact on group dynamics. It also provided an object lesson on the handling of power asymmetries.)

Location

Physical arrangements are of particular importance for any workshops whose participants come from different sides in a violent conflict. The venue must be neutral and accessible (in terms of transport, passports and visas) for all participants, which in war-torn areas may be difficult. In any case it is important to find a pleasant place with adequate comfort, privacy and food, light, adequately large and well-ventilated work rooms, with smaller areas for group work and plenty of wall space for displaying flipcharts. All this within the constraints of available budgets! If conditions cannot be ideally comfortable, the effects of physical discomfort need at least to be taken into account, and options for their management discussed with participants. Providing drinks and snacks helps keep up energy and maintain good temper.

If the venue is too close to the temptations of beach or town, it may be hard to secure participants' concentration. On the other hand, they are likely to expect some opportunities for relaxation, so accessible sites of interest or entertainment, or at least a free day and some provision for sightseeing, may be desirable.

This is a vital level of care which will not only optimise the quality of their participation and corresponding learning, but will model the respect and care which are seen as values for conflict transformation.

CONCLUSION

The world of workshops, small from outside, is vast once entered. Despite the length of this chapter, I am aware of having paid no more than cursory attention to most of its elements. Workshop facilitation is learned largely in action and each facilitator brings to it her or his own patchwork of theory and her or his own personality. Moreover, there are as many potential workshop designs as there are situations in which to work. What I hope to have done is to give some indication of the range of topics which may be relevant and some of the many possible ideas and means for addressing them with participants. I have also tried to draw attention to some of the challenges and pitfalls of working in these ways in situations of conflict. The workshop accounts which follow will illustrate how things can work out in practice.

Note: Training Manuals and Handbooks

Particularly relevant for work in violent conflicts are International Alert's *Resource Pack for Conflict Transformation* (1996, to which I contributed ideas and materials also presented here); the *Working with Conflict* manual produced by Responding to Conflict (Fisher *et al.*, 2000); Mitchell and Banks's *Handbook of Conflict Resolution*; *Working for Reconciliation: A Caritas Handbook* (Caritas, 1999) and the *Mediation and Facilitation Training Manual*, whose latest edition was published by the Mennonite Conciliation Service in 1997.

Examples of manuals useful in more local work are *The South African Handbook of Education for Peace*, published in 1992 by the Quaker Peace Centre, Cape Town; *Playing with Fire*, published jointly by Leaveners Experimental Arts Project (LEAP) Confronting Conflict and the National Youth Agency (UK) in 1995, and, with a more specific focus, Mediation UK's *Training Manual in Conflict Mediation Skills*, published in 1995.

These manuals and handbooks are also listed in the bibliography.

Appendix 4.1
Goss-Mayr Models for Analysis and Planning

ANALYSING THE SITUATION

In this model for analysis, the oppressive situation or injustice (in some circumstances more usefully defined loosely as a 'problem' – at any rate the thing needing to be changed) is depicted as an inverted pyramid, held in place only with the help of certain props or pillars, that is, by particular groups or sections of society which by their passivity, action or collusion support the status quo. Once the injustice has been defined, it is the task of the group doing the analysis to name these pillars which support it.

The example in Figure 4.1 is that of peasants living in the Algamar region of Brazil who, having no legal documents to prove their ownership, had been evicted from their family land by multinational companies. They defined the injustice as shown, including the wider definition of 'unjust rural structures', because their experience was part of a wider process of land confiscation. They named themselves as the first pillar supporting the injustice, since they had remained passive and submitted to the eviction. Then they named the other groups responsible: the landlord (the multinational company), the Church, whose hierarchy wished to align themselves with the rich and influential, the labour unions which had failed to act for them, the state army, acting for a government that supported the multi-national's acquisition of the land, the company's private militia, used to intimidate the peasants, the political parties which had

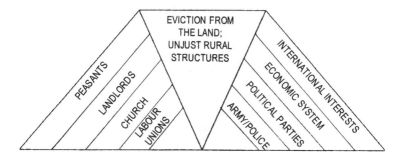

Figure 4.1 Analysing the situation

supported the government's policy or failed to speak up for historic land rights, the economic policy and system that preferred foreign investment to the needs and rights of the local people, and the international interests which exerted so much pressure on the internal affairs of Brazil.

CONSTRUCTIVE PROGRAMME

It is not enough, however, for a group to know what it wants to get rid of – though it is a very good starting point. They also need to know what they want in its place. They need a vision, goals to work for, a new way of doing things that can start immediately, so that the new is being built before the old is demolished. This idea was central to Gandhi's understanding of nonviolence. He called it a 'constructive programme'.

In the second diagram, shown in Figure 4.2, the peasants' programme for the future begins at the bottom with their own local victory, then goes on to a process for developing their own efficiency and economic power, then to their inclusion in the political processes of their country and thence to a national campaign for land reform.

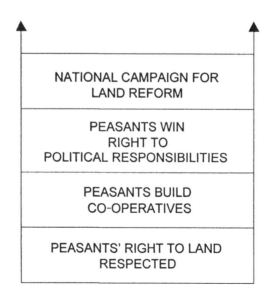

Figure 4.2 Constructive programme

BUILDING SUPPORT

Once the pillars that support the injustice have been identified, the group concerned can begin to consider how they can be removed or eroded. They must ask themselves why those they have named currently support the injustice and how they might be won over or persuaded to change. Realising they are at present small in number and relatively powerless, the group also needs to consider how to build support. In the third diagram, shown in Figure 4.3, the Algamar peasants, who belonged to one Christian base community, are depicted as the centre of a campaign which will grow, first to include their most natural allies, the other base communities in their region, then, with their help, the Church at other levels, the labour unions and the press (who at this stage can be expected to be interested). With this much wider backing, the political parties can be approached, and those in other countries who could bring

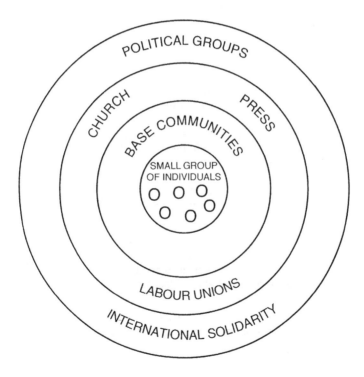

Figure 4.3 Building support

pressure to bear on multinational companies and on other governments, at the same time drawing attention to any repressive counter-measures taken in Brazil.

It can be seen that those who appear in the first diagram as props for the injustice in question can also reappear as potential allies in the movement for change. When such a transfer of allegiance takes place in reality, a considerable shift of power has been achieved.

Appendix 4.2
The Problem-solving 'Iceberg'

The diagram shown in Figure 4.4 encapsulates the process of problem-solving and the attitudes and behaviours on which it is ideally based. It has both the virtues and disadvantages of simplicity, providing a useful framework for discussion.

Figure 4.4 The problem-solving iceberg

The tip of the iceberg, showing above the waterline, represents the goal of the problem-solving approach to conflict resolution – an 'inclusive' solution, one which is acceptable to all parties, meeting at least the deepest needs and most urgent concerns of each. This is what we see – the outcome of the process. What is invisible, hidden below the waterline, is the major part of the iceberg, representing everything that needs to be in place or needs to be done in order for that goal to be achievable.

At the base is 'Affirmation', explained as 'self-respect and respect for others'. This includes recognition of the needs, rights and identity of each person or group involved in the conflict. In real life, this may be a hard requirement, and the respect which can be found initially may be grudging or purely notional, emanating more from a sense of necessity than from any positive impulse. Feelings may be running high and prejudice may be strong. It may require the efforts of mediators to make any kind of respect seem remotely possible and to generate enough trust for dialogue to begin. The parties may be driven to negotiation more by necessity than good will, but without some basic recognition and valuing, it will be hard, if not impossible, for them to take each other and the process of problem-solving seriously, or to communicate effectively, which are the first needs for problem-solving.

Communication, then, forms the next layer of the iceberg. Constructive communication is both a symbolic and practical expression of respect, whose goal is to build an understanding of the nature of the conflict and the needs and perceptions of the different parties to it. Ideally, constructive communication requires, as indicated on the diagram, both careful, empathic listening and clear but sensitive speaking, focused on the expression of needs and perceptions, rather than accusations. In practice, at least a modicum of listening, politeness and constructive thinking and speaking will be needed for progress to be made.

On the basis of respect and the opening of constructive communication, co-operation in the search for a mutually acceptable solution becomes possible. 'Co-operation' is therefore the heading for the next layer of the iceberg, and the basic processes of problem-solving are mentioned: reframing the conflict; shifting focus from positions to interests; generating options for contributing to a solution; evaluating and selecting from those options; reaching agreement on a settlement 'package' and how it is to be implemented. Since both analysis and imagination will be required in the problem-solving process, those two words are written down the sides of the iceberg.

Appendix 4.3
Needs and Fears Mapping

This exercise is a tool for both analysis and empathy. Although its analytical scope is not exhaustive, by including all parties' points of view in the definition of what is at issue and by acknowledging the needs and fears of each party in relation it, it helps reframe the conflict in question as a shared problem needing a common solution.

This exercise can be used by any group or individual wishing to deepen their understanding of a conflict. For instance, it can be used

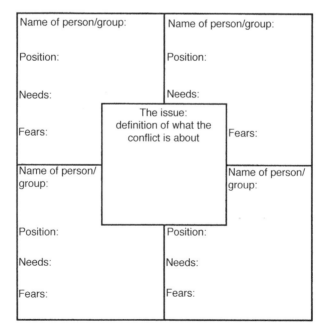

Figure 4.5 Needs and fears mapping

by potential mediators as a tool for research, or for organising existing knowledge about the parties and their motivations, and for 'getting alongside' them.

It may be used by one of the parties as a way of examining their own underlying feelings in relation to the conflict and attempting 'to put themselves in the shoes of the other', in order to see and feel things from their perspective in preparation for dialogue. It may also be used by all of the parties, separately or together (or both), with or without a mediator, as a means of reciprocal explanation and understanding. It can help them to hear and understand each other and constitute a first, important step in co-operation.

To focus on needs and fears can help to free those in conflict from fixed positions in relation to it and to focus instead on the interests which will need to be addressed in any future agreement. In addition, listing needs and fears often reveals the multiplicity of those interests and the need for several, or many, ingredients to be included in that agreement.

As an exercise in a training workshop, this form of conflict analysis helps participants to understand the practical use of needs theory as applied in problem-solving and provides them with a tool for future use. The instructions below outline the steps for the use of the exercise in training:

1. List the different parties to the conflict – all who have some stake or involvement in it. (In social or political conflicts, parties are often not homogeneous or united, but have factions, leaders and followers, core members and supporters, etc. It will then be necessary to identify those who are the most important to include.)
2. Write down, if you know them, the current positions of all the parties identified, that is, the demands they are making or the stated goal of their struggle.
3. Find a definition of the issue(s) or problem(s) which could be acceptable to all those involved – one which they could possibly work together to address. (This can be very difficult, since the issues may be many and disagreement over what the conflict is about may itself be part of the problem. It may be necessary to accept competing definitions, but the discussion will, in itself, be illuminating.)
4. Think what are (a) the needs, and (b) the fears of the different parties.

(Sometimes the fears that are listed may be just a negative way of expressing the needs, but often this question uncovers more deep-seated concerns that might otherwise be overlooked.)

These elements can be arranged diagrammatically on a flipchart, with the definition of the issue in the centre and the names of the parties, with their different positions, needs and fears, arranged around it, as shown in Figure 4.5.

The use of a rather simple method to analyse complex situations may appear counterproductive. In fact, as indicated above, the mere attempt to understand a conflict in terms of a small number of simple questions may highlight just how complex it is – the multiplicity of issues and subdivisions of parties, and the aspects of the conflict not addressed in this form of analysis and calling for further thinking, such as the different degrees of power of the parties in relation to the conflict and to each other. To recognise these complexities makes it possible to begin to address them, or to see what can be addressed – which is the purpose of such analysis. In whatever context it is used, this mapping exercise is thus the beginning rather than the end of the analytical process.

5 Transcending Culture: International Relations in Microcosm

This chapter will present the detailed record of a training workshop organised by an international and ecumenical church body as part of its programme for 'overcoming violence'.

The workshop, which lasted more than a week, was for people working for peace and justice in all parts of the world, especially in situations of conflict and violence. Its aim was to give them an opportunity to share experiences and reflect together, to discover new ways of seeing things and acquire some new tools and skills which would be of use in their own, different situations. It did not in itself constitute a 'conflict intervention' but was designed to build the capacity of individuals wishing to work for conflict transformation in a variety of situations, as well as the collective capacity of the world church body of which those individuals were members.

I was at the original planning meeting for the workshop and was invited to take on responsibility for its design and facilitation, finding such colleagues as seemed appropriate. I was clear that I needed to work with someone from the south, and through the good offices of common friends, succeeded in securing the partnership of a South African pastor from King Williamstown. He had worked for the Ecumenical Monitoring Programme in the run-up to the first post-apartheid elections in South Africa. He was also a key member of his local Peace Committee. Although he was an experienced trainer within his own country, he was diffident about working with an international group. I wrote to him to argue my case:

> I quite understand your hesitations. I feel the same. I think, 'Who am I to do that?' Then I think that it is the people themselves who come together who will bring the experience and the wisdom. What the trainers have to do is to provide a framework, be responsive, facilitate a process; and maybe we are able to do that, with each other's support.
>
> I have more reason than you to feel diffident. I come from a relatively very safe part of the world and in some ways live a very

comfortable life. That doesn't mean I have nothing to offer; but it does mean that as co-trainer or co-facilitator I should work with someone from a very different setting, with a kind of daily experience, and therefore perspective, that I lack.

I was relieved when he agreed to work with me. When we met to plan the workshop it soon felt as if we had been working together for years. He subsequently told me that this time we had, working together on the planning, had given him confidence both in me and in his ability to work with me.

The workshop (held just outside Geneva) was entitled 'Living with our Differences: Nonviolent Responses to Conflict'. The coming together of people from so many backgrounds and cultures provided us all with an opportunity to benefit from the insights gained from comparison and from the brief but intense experience of living in a cross-cultural community. Experiential learning is a strong thread in my account, in which common themes and values emerge as well as points of tension and difference. It also illustrates and discusses issues of co-facilitation and the role of facilitators, their exercise of power and responsibility, the pressures experienced by facilitators and the kinds of decisions they have to take in response to changing demands and moods within the group, and the changing dynamics, roles and power relations it embodies. The question of gender relations takes centre stage for a while. The account demonstrates both the benefits and the challenges of such broadly based, multi-cultural workshops. In this case cultural differences, though substantial, were at the same time transcended by a shared religious culture, which gave participants a common language and shared aspirations. The daily worship, which the participants organised and led, drew them together at a profound level.

Note the use of the stages diagram (Figure 2.1) as a framework for the workshop's agenda and an introductory mechanism to set out certain concepts and engage participants in thinking about their own conflicts. Note also the use of base groups as a vehicle for the distribu-tion of power and responsibility, and for evaluation and feedback. In this Christian context, in which ethics played a key role, the question of violence and nonviolence was unavoidable. At the same time, given the experiences of some of the participants, it was highly charged. The timing and manner of its introduction and handling are discussed in the account. The use of time was a bone of contention in this group, as it often is, illustrating the difficulties of balancing rest and work, responding to different preferences and priorities, and covering certain ground without over-packing the agenda.

The account is long and detailed. It was compiled soon after the workshop took place from my daily records and journal writings. I have decided to reproduce it more or less verbatim, with the idea that to give one such detailed account will ground the reader in the moment by moment reality of what these workshops are like in a way that generalised descriptions cannot. All names have been changed, as they have elsewhere in these chapters.

The workshop began uneasily, since the organisers had been unclear in their invitations as to whether the workshop was an introductory one or a 'training for trainers' workshop. In the event, two of the participants who came were already trainers, while others were at their first training ever (a salutary lesson about clarity of purpose). We decided to explain our dilemma to the two trainer-participants, being clear with them that a mistake had been made and that we needed to work on the basis of the workshop's original purpose as broad, introductory and inclusive. We would invite them to consider how they would cope with any difficulties this would present them with.

Having given the reader this introductory information, I shall reproduce the account from the first evening of the workshop. Extracts from my journal will be given in italics and any later commentary will be marked by square brackets. I have called my co-facilitator George.

Opening Session, Friday

Our first session, on the Friday evening, was devoted to an official welcome from the organiser, introduction of the participants and trainers (first in pairs and then by each other to the plenary) and some scene-setting for the workshop as a whole. I explained that we would be seeing faith and practice as inextricably bound together, with our daily worship an important part of our programme; that we wanted the group to become a community of learning, using participatory methods and working in an informal and relaxed atmosphere; that the agenda we had outlined was therefore intended to provide a framework for the sharing of experience as the basis for the development of understanding, skills and commitment; and that the sharing of spiritual insights and resources would be integral to that process. The direct and immediate experience of living together as a group would provide a challenge to us all, coming as we did from very different cultures and lifestyles, and doubtless with some very different perceptions of the world. We would hope to discover much that was universal, but also to recognise, respect and even celebrate our differences. In a quick brainstorm of words associated with conflict, painful associations at first predominated; but, with encouragement, participants produced many positive words as well. We wanted to establish that conflict is not only an inevitable but potentially a creative part of life. What matters is how we respond to or enter into it. We drew participants' attention to the assumption implied in the title and planning of the workshop, that we would be exploring ways of doing so nonviolently.

We apologised for the confusion over the nature and intentions of the workshop, explaining that it would not be 'training for trainers', though we hoped those who were already trainers would also learn from the whole experience.

George introduced the idea and composition of the base groups we had designed for support, evaluation and worship preparation, then went on to outline our planned daily timetable. This caused something approaching an uproar, being seen by many as overloaded. We re-explained our thinking, but also promised to reconsider, taking into account the many different preferences being voiced by participants.

(We did not explain at this stage, but might usefully have done so, that we had planned that our daily agendas would be interspersed with songs and games, partly to provide a change of dynamic: to help people relax after tense or difficult work, to wake up after lunch, to laugh after sadness; to meet in a non-cerebral way; occasionally to experience in symbolic or bodily ways things which had been matters of discussion, such as co-operative problem-solving in the game called 'knots' (where participants make a tight circle, close their eyes, reach out and find two other hands to hold, open their eyes and then try together to resolve the resulting tangled knot into one or two circles). I was afraid the bishop in the group might feel that 'children's games' were beneath his dignity, whereas in fact he loved them, which only goes to show, I suppose, that such judgements are hard to make and that sensitivity and prejudice are hard to distinguish.)

In our post-session evaluation, George and I agreed that things had gone well till we came to the revolt over the timetable. We felt we would have done better to describe the evenings as 'free but' (with optional activities) rather than as 'work but' (the right to opt out). We thought we had supported each other well, but that maybe we had fought back rather too hard over the agenda! Now we had to be prepared to make concessions. I realise now with what a conflictual attitude we had all begun, the 'experts' and generally powerful characters in the group wanting to flex and display their muscles, and George and I wanting to be clear about our role and responsibility – and therefore, at this stage, our authority. In my journal late that night I wrote:

> *I think George and I dealt with the turmoil in a mutually affirming way, also respecting the views of the group, wishing to give the voiceless a voice, as well as hear those who have already spoken. (This by asking for feedback from the base groups after further discussion.) We've made a complete redesign for tomorrow morning, to respect the process and check more carefully how people respond to our overview of the agenda.*

Saturday

On the Saturday, the first full working day, when we had gathered in silence, I began by introducing the 'stages' diagram which we had used as the basis for our agenda outline. This diagram and explanation were received with enthusiasm, and many participants were quickly able to identify with the stages described, relating it to their own experience, locating themselves in the present at different points on the 'snake'. They made suggestions for additions and modifications to the diagram (which I had drawn on the white board) and I wrote them in.

In the second part of the morning, George invited participants to suggest, on the basis of some initial discussion in 'buzz groups', what we all needed from each other to become a working community. From these ideas we agreed some mutual commitments. After a game, people then went off to do some group work, sharing their experiences of violence in its different manifestations. The learning drawn from this process was reported back to the plenary, after a two-hour lunch break, and I offered a definition of violence as whatever is done by choice which harms, oppresses or destroys, or which prevents the fulfilment of human (and other?) potential. I also introduced Johan Galtung's (1990) idea of the triangle of violence: direct, structural and cultural.

George went on to categorise options for responding to violence: remaining passive; reacting with counter-violence; or responding creatively, nonviolently. He asked participants to return to their groups to consider their own experiences of these three ways of responding and to try and identify some common characteristics of the third way.

After plenary feedback from that group work, we briefly evaluated the day, asking participants what they had liked and not liked, and what ideas or wants they had for future sessions. All had enjoyed the group work and some would have liked more time for it; the 'snake' diagram had gone down well and some would have liked more time for that too; the facilitation and participation had been felt to be good and the interpretation was appreciated. The heat had been very trying but people thought we had all coped well. Some had enjoyed the long siesta, others had found it too long. Suggestions for the future mostly related to this last topic, though there was also a request for more theology and a proposal, vigorously countered, that the clergy should organise all the worship!

Each day closed with worship focused on a topic relevant to the day's work, so at this point we went off to the chapel, with the reminder that base groups were to meet after supper for further evaluation and reflection, and to report to us with feedback and suggestions. The person who was to have prepared this first evening was unwell, but other members of her base group put something together in a very satisfactory way.

Later, as we waited for the base group reports, George and I reflected that things seemed to be 'going OK'. There had been much positive feedback, especially on the first session and the 'snake', which seemed to have drawn in even those who had seemed inclined to be antagonistic. The reports from the base groups confirmed this impression. Participants felt we had worked hard to respond to their needs (one or two were impatient with all the demands and wanted us just to proceed according to our own plans), and most were content with the way the timetable had been arranged. Two individuals (who remained vocal throughout) wanted all afternoons free for sightseeing, but they were strongly countered by others who wanted to work in the day and appreciated time for informal exchange in the evenings. The desire to have more time for everything was a recurrent theme throughout the seminar, and on this occasion one group expressed dissatisfaction with the amount of time allowed for plenary discussion. In fact our agenda had been eaten into by wrangling over time allocation, and we had been reluctant to lose content and had therefore perhaps tried to include too much, here and elsewhere. Members of one group had felt that more equal participation in plenaries should be encouraged, also that it would help if some who were failing to do so could share real examples from personal experience, not just theories and generalisations.

My journal entry for that night reads as follows:

We've taken all the feedback from the base groups (work well and thoughtfully done – with some fairly inconsiderate agenda-pushing mixed in) and we're trying to acknowledge all needs and wants and meet the most urgent, bearing in mind the needs of the seminar and its content. George and I have felt fairly overwhelmed by what has at times seemed like a clamour, but have received also good affirmation, thanks and care, and good-quality, honest observations from some.

Sunday

The following morning, Sunday, after silence and a song, I delivered, as George and I had agreed, a report on the feedback we had received from the working groups, thanking them for the quality of the work they had done and the helpful way they had reported it to us. I then explained the decisions we had reached and why: to offer an agenda with substantial chunks of group work preceded by plenaries; to stay with the timetable of the former day; and see, after one more evening's feedback, whether we could stay with that or would need to alternate longer afternoon siestas with free evenings; and to offer the whole day free on the Wednesday, cancelling the evening session which had been scheduled. I explained that this was a concession particularly to those who had expressed a wish for more time for sightseeing, and that it was the only concession we saw fit to make. Most of the group wanted to focus on the work in hand. The church organisation had invited and paid for people on that basis and it was our responsibility to maintain that focus. Relaxation and informal exchange and friendship enhanced the work done in sessions, but a balance had to be maintained, and we hoped that what we were now suggesting achieved that balance.

This explanation and these decisions, which we asked the group to accept, were indeed accepted, I think with much relief by most, if not all.

We proceeded with a continuation in threes and fours of the previous afternoon's reflection on the characteristics of the nonviolent response. Many ideas emerged from this discussion, and some questions. In the subsequent group work and plenary discussion on the resources and inspiration for nonviolence in people's own faith and experience, these ideas and questions were further elaborated. The faith basis for nonviolence was clear to all, and the disastrous consequences of violence, but several remained doubtful as to the efficacy of nonviolence.

In the afternoon we went on to look at the methods of nonviolent empowerment: at group formation and the work of analysis which will need to precede action. First a model was offered for defining the injustice or violence which the group wishes to remove, and identifying the things – or the actual people – which support it or kept it in place. We used the example of a peasants' land struggle in Brazil (see Appendix 4.1, page 123). Then we presented a model for planning a campaign to build solidarity, and one for setting goals for alternative processes and structures. Next, participants were divided into groups to share accounts of situations of violence they knew, and choose one for a case study, using the models which had been presented. This work took us to the end of the session, which closed with a song and a brief evaluation.

The work content was felt to be evolving well and the time spent in groups had been greatly appreciated, both for the opportunity it had afforded people to get to know each other and for the chance to exchange experiences and insights and analyse participants' own situations. There was a feeling that some people had tended to dominate plenary discussions and regret that there had been no opportunity to discuss passive and violent responses to violence, as well as the nonviolent way.

After the evening's worship and supper, George and I once more reviewed the day as we waited for the base groups to report. Overall we felt very happy. The group work we considered to have been excellent. We agreed to remind people to rotate the tasks of group facilitation and reporting.

What emerged from the base groups confirmed the points made in the earlier evaluation. Our responsiveness as facilitators had been appreciated, as well as our clear assumption of our role. The daily pattern was acceptable to almost everyone, and all were willing to go with it. The group work had given people a real chance to share feelings. Practical proposals were made about cold drinks, evening refreshments and workshop reporting, and an offer was made by one participant to present a case study of a campaign he personally had been involved in, using the day's models, in the following morning's plenary.

At some point during the day I had written in my journal:

I realise how important communication is in fleshing out respect, as well as being an outcome of it. It's a great help here being able to talk with most people: the importance of language.

At bedtime I wrote:

We've had a good day. Patience, intense work, intelligent weighing of differing needs, a will to balance responsiveness to wants with the requirements of the task and fair expectations of participants led this morning to a caring but assertive set of statements of intention which were accepted.

The difficulties of remaining constant in nonviolence (that is, sticking to it) are keenly felt by some participants. How can we (George and I) acknowledge that – give it time – while pursuing our agreed goal of focusing on the nonviolent option?

Monday

The following morning, the Monday, after summarising the base group feedback from the previous evening and explaining our thinking about the coming day, we took the case study reports from the groups which had worked the previous afternoon, using the three models for empowerment. I told the story of one particular action taken by the Brazilian peasants whose campaign had been used as our original example: an action of outreach to the militia. Then the participant from the Philippines, who had offered the previous evening to do so, told the story of a group of village women he knew who had used the models to analyse their own case of having their fishing rights removed, and plan a campaign for their restoration. This first-hand account of a lived example made exciting listening and wholly engaged the group. (I now see that if I had had time to consult with this participant before the previous day, this example would have been the one to use in the first place.) Key ideas in this session on self-empowerment by oppressed groups were the need for constant reaching out to the other and dialogue, for building solidarity and trying to shift the opposition; also the need to calculate costs and risks and to be tirelessly persistent. After coffee, participants returned to their groups to continue to work on their case studies – one on terrorism in Egypt, one on the continuing deforestation of Ethiopia, and one on the 500 years of colonialism and the expropriation of land in Latin America. The task now, building on the analysis, was to set objectives and devise a plan of appropriate action for the initial stages of a campaign.

After lunch, siesta (or for some a shopping expedition) and a game, the groups went on to select a particular episode in one of their planned actions to test out in a role-play, to gain some insight into the human dynamics of such action (including the feelings of the opposition) and a sense of the skills and resources that would be required by the nonviolent activists. The groups returned to plenary too late to report on their action plans and their role-plays, so reports were held over till the following morning and we went straight into evaluation. The participants had been really engaged in the group work and were generally tired and content. The work of our inter-preters which had made the group work possible was greatly appreciated. (One group at least had included language difference as a feature in their role-play, both as a practical expedient at the time and because it was real for their chosen situation, where indigenous people were lobbying the ruling ethnic group in Guatemala.)

George and I reflected later that it would have been useful to have spent more time at the beginning of the workshop on group dynamics and helpful and unhelpful roles and behaviours of indi-viduals in groups. (I don't remember what exactly gave rise to that thought.) We noted that it would have helped the groups, when they were making their action plans and preparing for the role-plays, if we had asked them to be sure they were clear who they were in the chosen scenario. We were greatly relieved and really delighted that the group which had had greatest difficulty at the stage of analysis (the Latin America group, which had had to use interpreters throughout) had really 'got into it' when it came to planning for action, and had done a wonderful role-play – incorporating the language problem (see above). All the groups had worked through the entire task, experiencing their own difficult patches and blossoming at different points, and we had gone with the flow, supporting where necessary and stepping right out of things at other times. Our instructions had not always been closely followed, but had provided a framework for real engagement with real issues.

Feedback from the base groups supported this assessment. The extended group work had been greatly appreciated – the chance to follow something right through. There was some eagerness to be moving on to the 'conflict resolution' part of our programme and some concern that the programme for the following day now looked overloaded. Someone offered to present something in relation to 'dealing with stereotypes'. Satisfaction was expressed that our evening worship was integrated with the content of our work. A deepening sense of community was noted within the whole group. In my journal that night I wrote:

I think again in our facilitation today we respected our own agenda and the energy and direction of the groups.

One interesting thing said by a Zambian to a South African (in an evening conversation over wine) was that it was not acceptable in his culture ever to remind someone of past favours or to look for gratitude. This was in reference to the economic sacrifices made by Zambia, and ordinary Zambians, in solidarity with the liberation struggle in South Africa. Zambians clearly feel patronised by South Africans, which from my observations they are. But the South Africans also clearly recognised the costly support they were given. They sang for Nelson (the Zambian) the song of gratitude sung by ANC [African National Congress] people leaving Zambia to return home.

I noted in my journal that the Burundian bishop, who had seemed so far a little uneasy, had relaxed, made some very good jokes and joined with enthusiasm in the games we had been playing. As the week went on, we were ever more aware of the pain and anxiety he carried for what was going on in Burundi. If we had been more aware in the first place, we could perhaps have supported him better from the start.

Tuesday

On the Tuesday morning George and I had decided to make space for the airing of doubts and difficulties about nonviolence, the debate which some had been wanting to happen. It seemed to us to make sense first to have a good look at the philosophy and methods of nonviolence. Now that we were coming to the end of this section on empowerment, we needed to come back to the unfinished business of doubts and obstacles before moving on. So after reports from the groups on their previous day's work, we invited participants to think how the work they had been doing related to their own situations, and to consider what were the obstacles to following 'Jesse's third way'; what kept people in passivity or drove them to counter-violence. We invited them to share their thinking first with two or three others.

The plenary discussion that followed was long and very heavy, taking the rest of the morning. The pain and despair in the group were almost palpable and we seemed to be going down and down. I felt very diffident about intervening, but there came a point when I felt we needed some reference point outside our present pit, and spoke in recognition of the pain, but also for hope, for what Adolpho Perez Esquivel (Nobel Prize Winner from Argentina) calls 'relentless persistence', for starting small, for what Luis Aguirre (a Uruguayan colleague) calls the 'seamless garment' of the worldwide movement for transformation. I think I chose the right moment, and we climbed up and out, ready to move on. What had taken place was not a debate, but an outpouring, not a challenge to faith, but an expression of doubt, frustration and grief, and through this process we had been drawn more closely together in our shared hopes and aspirations.

In the afternoon we opened with a good rowdy game, then returned to the 'snake'. We had reached the stage described as 'conflict resolution' and formed new groups of five or six to look at the questions, 'What are our insights and values as people of faith for living with our differences? What are our principles for coexistence?' These groups reported back after tea, having evidently shared and generated some profound understandings at many levels. We then brainstormed ideas about what in our experience makes things better or worse in a conflict: helps and hindrances in the moment. The list produced was somewhat overloaded with abstract nouns and short on specific behaviours. This reflected a recurrent tendency in the group to feel more at ease with theories and generalisations than with immediate and specific experience.

To round off this session I drew and explained the 'iceberg' model for conflict resolution (Figure 4.4), which begins from a base of respect, on that builds communication, and through communication works towards the co-operative task of generating and selecting options for an inclusive solution to the conflict.

The closing evaluation was very positive. Participants had found it moving to share their feelings in the morning's session: their 'inner selves' as one person put it. It had been a 'highlight'. And one commented, 'It's OK – it's a group.' Later I wrote in my journal:

At the end of the morning there was a general feeling of achievement, having completed the first part of our work – the first section of the stages diagram. The afternoon seemed fine, but I was too tired to judge any more. The evaluation of the afternoon was a bit sketchy, as a good few had gone to prepare outdoor worship – which was very moving. Base groups failed to meet – which was fine. We all felt on holiday. [The next day was to be free.]

Wednesday

The Wednesday was, for the participants, an opportunity to visit the headquarters of the church organisation, have a nice lunch and see something of Geneva. For George and me it was a chance to recuperate, just *be* together, and take stock and make plans for the rest of the week. Our agreed outline agenda and the constant evaluation and feedback process we had devised made this as easy as it could be. The planning took many hours, but was well interspersed with personal conversation of all kinds. George astonished me with accounts of things he did as part of his Xhosa culture – like slaughtering an animal to bring an end to a run of bad luck. This for him sat perfectly comfortably with his Christianity. It made me wonder what things I did and took for granted that others would see as at odds with my proclaimed (or assumed) beliefs. It seemed strange that we should feel so much at ease with each other in that place and work, when we carried such different worlds on our backs. We also exchanged accounts of our parents' deaths, a very intimate thing to do. We shared many of the same assumptions and feelings, though the scenarios and events were also very different. In all of this the thing which struck me and touched me was that we were able to talk quite freely, being as open about surprise and difference as we were about recognition and sameness.

Thursday

On the Thursday morning we looked in more detail at the 'iceberg' model (Figure 4.4) and then went on to do some listening exercises. We used a representation of a mouth and an ear and the space in between to invite thinking about the complexities of communication which arise from the moods, expectations, attitudes and capacities of speakers and listeners and the contexts in which they find themselves.

After the break and a discussion of the characteristics of assertive speaking, participants were asked to think of a situation in which they had been passive or aggressive rather than assertive, and to describe that situation to their partner, asking them to be 'the other' and trying out in a mini role-play a new, assertive approach.

Leonardo, a Spanish-speaker from the Dominican Republic, had provided us with an excellent role model the previous afternoon, when he had come to the microphone at the end of the evaluation and said, 'When I get up and speak and everyone has to run for their headsets, I feel discriminated against, because I have to wear mine all the time to hear you, and I want to be able to speak without having to wait for you to be ready.' I had used this as an example when introducing the idea of assertive speaking. Eli, a Finnish participant, offered us a neat formula for helpful communication: Smile, Open, Forward, Touch, Eye-contact, Nod (whose first letters spell 'SOFTEN'). Some people were pleased with this, but when we questioned the whole group it emerged that these guidelines in many cases were culturally determined and could be taken as universal only in spirit, not in the particular – a useful learning.

Before lunch we had time for a brief discussion about the difficulties of handling strong emotions in order to be appropriately assertive as we would wish.

In the first afternoon session, thinking of stereotyping as one obstacle to good communication, we looked at the question of identity and belonging, the different groups we were part of and things about those groups that we were either proud of or ashamed of. We also recalled times when we had been, on the one hand, victims of discrimination and, on the other hand, guilty of discrimination ourselves. We spoke about the need to be critical of our own cultures, as well as valuing them, and able to respect others and their identities while at the same time being able to take issue with a particular approach or behaviour.

After tea we introduced the idea of 'reframing', or finding new approaches to relationships and problems. One form of reframing needed is a shift from seeing a problem as being caused or suffered by one party, with one alone able to solve it, to seeing it as being the affair of two and often several parties who all need to contribute to a solution and be included in it. In 'needs and fears mapping' the parties are identified and their different needs and fears listed – a method of gaining some insight into the way each relates to the problem. This George introduced, using by way of illustration a case from his own experience of local government in South Africa.

Before going into the final evaluation I told the group that the next morning they would be trying out this 'needs and fears mapping' on their own case studies. I reported that one or two women had said to me that it would be good to have the opportunity to work in an all-women group at some point and asked whether this might seem an appropriate opportunity for that to happen. At this point 'the shit hit the fan'. Pain, indignation and incredulity were expressed by several men in quick succession. I asked for one of the women whose idea it had been to explain the request, which one of them did most gently, caringly and clearly. This helped some to be less defensive, but there was still much leaping to the mike by men – in some cases to complain that it was for the women, not the men, to speak and choose. Three other women did speak, but very briefly: two to support the idea of an opportunity to work together as women, and one to say that she herself did not have such a wish, although she had no objection to others doing so.

One of the South African men was very angry, not, he said, because he had any objection to the idea, but because of the way it had been brought to the group, by me, as he saw it. This was puzzling, since I had explained that I had been approached by women participants with the request, and was presenting it on their behalf. He explained that he couldn't see why the idea hadn't been raised in the base groups. So really what I think he was complaining about was the lack of openness at that level, rather than what I at first heard, which was that I was using my role as facilitator to push my own agenda. I suppose these two explanations are not mutually exclusive. He possibly saw it all as a plot between me and one or two individuals. Although that seems to me an unduly negative interpretation, since I was open and exact with the group about how I came to be making the proposal, I can see that this was aside from the regular procedure of individuals making proposals to the facilitators via the base groups. Had I grasped clearly what (I think) he was saying at the time, we could have discussed whether this was improper and why it had happened that way – which would have been interesting and might have revealed something about the dynamics, particularly male–female dynamics, in the base groups. As it was, I apologised for any lack of clarity in my explanation and the participant in question accepted my apology but still seemed very angry. I and others tried to talk to him at intervals during the next two days, but he remained angry and distant and kept saying it 'was over' and that there was 'nothing to talk about'. Then suddenly he was back with us, fully engaged and very helpful. Perhaps he had been dealing with problems of his own. It had been clear from the start that he was carrying a great deal of pain and anger from his own experiences in South Africa.

Before the session closed we reached agreement that if there were enough women who would like to work in a women's group the next day, they should do that. In the evaluation that followed, participants noted that it had been good for us to recognise that we had some unresolved problems ourselves and had had to handle our own conflict. The men had been affirmed within the process (I'd have liked some more real affirmation of the women) and we had a real sense of being a group, continuing to get to know each other better. On the negative side, it was noted that in the debate men had spoken more than women, that men had resisted the idea that women should have the opportunity to meet as women, and, conversely, that women had sought to discriminate between males and females.

There was also a complaint that not enough time had been allowed for plenary reports from group work, which showed that work was not taken seriously and that time management seemed more important than the depth of discussion. Here I interposed that this imputation of motivation – or lack of it – to the facilitators was out of step with the guiding principles for assertive speaking discussed earlier in the day, but acknowledged, and acknowledge now as I write, that we clearly had (mis)managed time in such a way as to engender these feelings, losing touch with the needs felt by some members of the group. Maybe, with hindsight, we should have jettisoned some input; maybe I should have screened out Eli's SOFTEN communication formula (see above) and run the risk of seeming to disrespect her and her group, which had proposed the inclusion of her contribution; maybe we should have taken time from another day. But what we had been striving above all else to do was to take advice which had already come to us from one of the base groups, to give plenty of time to the final 're-entry' phase of the workshop – advice which coincided with our own thinking – and therefore condense things, regrettably, at this stage in order to take pressure off time at the end.

We left for the evening worship all feeling, I think, tired and emotional. The groups that met for evaluation that evening were not the usual base groups, but regional groups that also had the task of preparing contributions for a 'cultural evening' prepared by one of the base groups. The feedback these regional groups gave us included the view from one quarter that I was very democratic and that this made trouble for me. Consensus was difficult to achieve and I would do well to be more dictatorial. This same group expressed its appreciation of the facilitators' role and comments, but felt that the contributions of both facilitators and other participants in plenary discussions on group reports were sometimes too long – we could all be briefer. And they made the significant observation that some of the more meaningful things shared in small groups people were not willing to repeat in plenary, so that the true depth of group discussions was not mirrored in the plenary sessions which followed.

The men in the Africa group, according to their report, had found the intensity of the work done in twos and threes difficult, and would have preferred to spend more time in plenary. (One of the things most of the women found difficult about many of the men was their unwillingness to stop generalising and theorising and share real personal experience.) The group also commented that the confusion in the original workshop invitations about the purpose of the workshop had created difficulties for them as participants and us as facilitators. They had asked themselves what they would have gained by the end of the workshop, in exchange for all the experiences they had shared. (I greatly appreciated the honesty of this report – felt honoured by it. I shared their frustration of course, but can also say that this was a kind of 'bottoming out' phase, and that by the end of the workshop these people, despite their justified criticisms, were glad to have been there.)

The Asia group had liked the day in general, but reported some dissatisfaction that the women's issues had been taken so seriously when other requests had not been met. Here we detected a harking back to one participant's regret that we had not spent more time on introductions – and he had missed the first round anyway, by arriving late – and that they had not had advance written information about participants. We decided to ask each participant to write a paragraph about her/himself to be sent out with a corrected address list after the workshop.

In my diary that night I wrote:

Great, happy party tonight, after all the grief. Only Leicester (from South Africa) missing and Sarah (the main speaker for the women's group proposal) still shaken.

We had processed our conflict sufficiently for the party to be wonderfully timely. We all relaxed and let go and revelled in each other's company and the richness of the group, and ate and drank and forgot, or at least set aside, the burdens we carried – even the threatened war in Burundi. The Asians performed a macabre skit about the disposal of a corpse and I was afraid it would be too near the bone for the bishop, but he entered into the spirit of things and laughed a lot and really seemed to let go of his troubles for the evening.

Friday

The next morning, after thanks to the party organisers, and our usual report on base group feedback and agenda review, we collected ideas on third-party roles in conflict, both positive and negative. We focused on the mediator role, and asked for examples from participants' own cultures: who might perform the role and how they would go about it. George then summarised the steps which seemed common to mediation in all cultures and got the group to brainstorm the qualities and functions of a good mediator.

After the break, four groups, including one of women only, worked together on their own chosen case of conflict, mapping the needs and fears of the different parties to the conflict, then devising a role-play in which some kind of mediation was attempted. This group work we allowed to take its own time, and in the event it lasted through most of the rest of the day, with reporting in the final session. We had invited the groups to consider the option of taking as their case study our own conflict of the previous day. The only group which seriously considered this option was the women's group, and in the event they decided to examine instead the injustices against women in their own countries, mapping their needs and fears. They produced a very substantial and sobering report, and never got to the role-play. In other words, they deviated from the set task and used some of what they had been offered for their own purposes – which seemed fine to George and me.

Another group had reverted to a previous case (the Philippino fisherwomen) and re-analysed it, using needs and fears mapping to report on the findings of their role-played mediation rather than to inform it – which also seemed fine. More regrettable was the plight of the group which had failed to get beyond a rather general and inconclusive discussion, on account, they said, of the lack of adequate interpretation and sporadic absences of group members. Only one group followed our proposal in a thorough and faithful way – and they were (or Heinz was) a little indignant that others had failed to fulfil the allotted task. George and I were delighted that the participants had little by little taken more responsibility for the running of things and had used the day in ways which were useful to them. We were not expecting uniformity and had supported the groups' choices as we went around.

Apart from the disappointment of the group which had been hampered by inadequate interpretation, the closing evaluation of the day was entirely positive. Some of the things said were an expression of the group's relief that we had weathered the storm over the women's group issue – that the feared thing had happened and that in the event it had not destroyed our unity. Our facilitation was appreciated and the ample and flexible time given for the group work and reporting back was gratefully acknowledged. There was another flow of gratitude for the great pleasure of the previous night's party, where all had belonged and participated – a proof of the 'culture of peace'. The poem by David (from Ethiopia), written and read in stages while the party was in progress was seen as having captured the spirit of the evening. Early the next day I wrote:

> *Working with George has been the great blessing of all this. He's not into self-doubt and self-blame. If we clearly made a mistake he'll say so, but from the point of view of someone who doesn't think that's a big deal – just part of being human: a very good lesson for me.*
>
> *Yesterday felt very relaxed and feedback was positive. People tried to be affirming of the women's group work (despite inappropriate laughter and profound incomprehension in some ways from some people) and the analysis the women produced – and instances – was powerful. Today again we're very relaxed. I was saying to George that little by little power has shifted from us to the group, though we still provide the frame. The group has grown into being progressively more of an entity able to take power productively.*

We had asked the base groups to give us some idea of their priorities for the use of our last full day together and on this, as in enthusiasm for the day we had just enjoyed, there was unity: we should proceed to the question of reconciliation and forgiveness, and work, in these final stages, in regional groups. This George and I had planned to suggest for the 're-entry' session, but now decided to propose for the whole day.

Saturday

We began the Saturday with some first farewells (sad), then explained the day. The first topic for the regional groups was as follows:

From hurt to reconciliation: what does it take, for ourselves and others? What are the ingredients and steps needed, both internal and external (within ourselves and out in society) from the point of view of our faith and our experience?

The plenary session in which we heard the reports from the four regional groups (Africa, Asia, Europe and Latin America, with the remaining person from the Middle East opting to work with the Asians) lasted until lunch. A few would have liked to discuss the reports – particularly the differences between them – in plenary. Others, wishing to preserve the regional session planned for the afternoon, preferred to continue the discussion, as appropriate, in regional groups. Seeing that this procedural debate could last indefinitely, George and I conferred and then offered our judgement: that the regional groups should reconvene immediately after lunch. This decision was accepted by all, with much relief. (Also in this session I had finally reprimanded Heinz for talking to his neighbour while someone else had the mike.) Later I noted again,

The group has, little by little, started almost running itself, power and responsibility having largely changed hands and we are facilitating very lightly. Ironically, we have also twice today (over how to continue the discussion on reconciliation and over the use of a room) made decisions for the group in a direct way for the first time and they've been glad. They're confident now in their own power and accepted our decision-making happily, seeing it as a service.

The first task for the groups in the afternoon was therefore to discuss the different reports from the morning, then to locate themselves on the 'snake' once more – and of course that positioning depended on the specific situations of different countries and the particular work of individuals. Then they were asked to consider how they saw their task now – what positive initiatives had already been undertaken and whether they had any new insights into their potential role.

The session's work was intense – especially in the large Africa group, which had had a tough but productive morning on the subject of forgiveness and now produced a most impressive report. But the others also found the time useful for drawing threads together and reaching conclusions. The closing evaluation confirmed the sense of satisfaction everyone had found in these regional sessions and the real substance of this final work together. Only one of the Latin Americans felt sad that, as he saw it, his region was peripheral in world opinion these days, and that this had been reflected in the workshop. (I think the language barrier had a lot to do with this feeling, and it remains a question whether enough had been done, in the circumstances, to overcome it. The Latin American group had certainly been isolated in many ways and at many times.) Leicester (the South African who had been so angry), whose re-engagement had been such a gift to the group, expressed his satisfaction that the whole group had been willing to work together till the end, and another participant was grateful for the reassurance of having reached the final stage of the conflict-resolution process and concluded that real solutions were possible.

The base groups did not meet that evening. We all went to see the annual firework display in Geneva.

Sunday

After our usual silence and song on the final Sunday morning, we played a game and then got down to the business of our final evaluation. The first question, considered in groups of three or four, was, 'Living with our differences: how did we do it – with our own differences of race, gender, culture, personality and regional perspective?' Then, working in the same groups, participants were asked to evaluate the content and process of our work, the way the group as a whole had engaged with it and the way they as individuals had engaged with it.

In this evaluation, it became clear that the learning which came from our own differences and the way we had lived with them had been of great importance to all the participants. They had 'learned a lot *about* other cultures, personalities and mentalities, different perceptions and understandings', but also experienced *directly* the difficulties and the rewards of being part of such a disparate community, having 'learned to live with our differences as a group, and how to work together – in group work and through the processes we learned for conflict resolution'. Another group said that we had all 'gone through the process' of empowerment and conflict resolution. We had begun with a 'single goal' and started by trying to find what was common to us all, but through the work and the living had found and affirmed our differences.

Recognising these differences had, in the words of another group, enriched us. We had learned to appreciate them, to respect each other, to be considerate and where necessary to compromise. We had also 'broadened our understanding of different situations in different countries through open sharing'. Yet another group described our achievement as 'having become a family here together. We had conflicts over contents and about the women's group, but we coped. We developed strong personal relationships. We found that coming from different cultures, races and ideologies doesn't prevent people from living together.' And the Latin American group, which had suffered so much isolation because of language, said nonetheless that our 'goal had been attained'. In spite of all our differences we had been able to be a model of coexistence, recognising our differences, accepting and discussing them to our mutual enrichment. Respect had prevailed. This had confirmed us in our commitment to the promotion of a nonviolent culture, a 'culture of life'.

It is clear to me from these evaluations that the learning that comes from the sharing of information and reflection, and the learning that comes from the experience of the group dynamics – which also involves the planned and explicit learning processes of the workshop – are not separate in reality, although they can be separately named and to some extent separately discussed. As the Latin American group put it, 'The methodology and the process dynamics were very important in bringing us to an understanding of each other.' Another group felt that what they saw as a concentration on input and skills in the first few days had at that stage stifled debate on issues that needed to be grappled with – thereby, presumably, hampering the healthy processing of differences. My own assessment is that we held them in suspension until people were able to cope with them more maturely.

The programmed content of the workshop was generally considered good and relevant, enriched as it was by cultural exchange, but also 'packed'. This was certainly a source of pressure and friction at times, and doubtless it limited potential learning.

On the other hand, the variety in the programme had made for enjoyment and full personal engagement, and the whole process was considered to have been participatory and open. The use of the base groups, and the flexibility and responsiveness of the facilitators, were appreciated. One group at least felt that time management had been good. The pack of written materials provided by an international nonviolence organisation was seen as an excellent resource for further learning and as a tool for participants who planned future training work. Our worship, in which we had shared at the deepest level our differences and our unity, was seen by everyone to have played a key role in our growing and learning together – along with our cultural evening and all our singing and playing, our talking and laughter. Through it all we had, as one group put it, 'internalised' the idea of living with differences. We would go back prepared to 'do' it. As someone else said, 'Peace is the way', and our workshop had been part of the peace process. We closed with a final act of worship, held under the trees outside.

My final evaluation with George was, despite our exhaustion, positive. Most of the feedback from participants had been positive and almost all the regrets that had been expressed related in one way or another to shortage of time. Since different people put priority on different things, we recognised it was impossible to please all of the participants all of the time. We felt the base group process for feedback, in addition to the daily end-of-session evaluation, was an excellent way of respecting the feelings and wants of participants, as well as our function as facilitators.

CONCLUDING REFLECTIONS

This experience of facilitation in a group of such powerful and varied individuals with so many potential sources of conflict, provided me with an excellent context for self-examination and learning as a facilitator. The power struggles that went on between facilitators and participants did not have a clear North–South dimension, though maybe that dimension was present, to some degree, in the discussion of nonviolence and the row about gender relations. This was a group where there were so many differences that lines of division did not, on the whole, form clearly, and George and I between us 'represented' both South and North. This complementarity, both real and symbolic, and the strong and easy rapport between us, was essential to our ability to work effectively for the group.

The overriding significance of this workshop, for me and it seemed for others, was that it brought together people of so many cultures in one experience and one discussion, that the experience was, for all its difficulties, a positive one, and that we seemed to have a discussion that had meaning for everyone, in spite of linguistic and cultural barriers. One of the participants wrote a report for the organisers which described the events in a way that confirmed my impressions and the accuracy of my recording. What interested me most, however, was the way in which she used the stages diagram to track the group's own journey, from a situation in which all kinds of conflicts lay beneath the surface, and power was an issue between us, through confrontation, to a kind of resolution, and a commitment to work together to maintain the community we had built. This was an apparent success in experiential learning (one which has not always been matched in subsequent workshops – see Chapter 6).

Workshops like this one bring me up against my own unease as a Western trainer working across cultures. I was helped by a passage from *Lost in Translation*, Eva Hoffman's profound account of the personal impact of migration from one culture and one language to another. She concludes that there is an essence within the individual which lies beyond culture (1991: 276):

> This is the point to which I have tried to triangulate, this private place, this unassimilable part of myself. We all need to find this place in order to know that we exist not only within culture but also outside it.

In my journal I wrote:

> *Is it possible that one of the most powerful and important things that training can do is to help participants discover that place for themselves? Is it also possible that communication at a profound level can take place between people even of very different cultures, if they can speak or otherwise communicate from and to that place? More specifically, can respect, if it comes from the heart, make itself felt even without words, or in spite of the wrong ones, or other cultural blunders? Is it something beyond concepts and words and world views? I ask these questions not to excuse wanton ignorance or carelessness, or the failure to look for local partners to work with when one is working outside of one's own cultural context; only to express the profound but tentative hope, or still more tentative belief, that there really is such a thing as 'our common humanity', which can be felt, and which can generate respect and make its communication possible.*
>
> *Cross-cultural (and counter-cultural) training for conflict transform-ation is bound to involve conflict at some level – the uncomfortable effects of one way of seeing or doing things clashing with another. A facilitator should not be afraid of conflict per se – including cultural conflict; but she or he should be aware of the impact of power relations and take care not to abuse the 'trainer' position. For me, so far in my research, 'respect' holds good, both as a core theme and value for those*

wishing to approach conflict nonviolently, and as a focal point for cultural differences and dilemmas. It also seems the litmus test for what constitutes acceptable cross-cultural training.

Many years later, and with a good deal of cross-cultural work behind me, I still hold to that hypothesis.

6 North–South Relations: A Pan-African Workshop

This chapter presents the account of a pan-African workshop for women trainers, held in Zimbabwe. It brings into sharp focus the issue of post-colonial relationships while at the same time demonstrating the cultural transferability of conceptual frameworks and tools and of the workshop approach itself. It also demonstrates the complexities of group dynamics, the roles of individuals within groups and (by default) the need for co-facilitation. The role of language is touched on – its power to unite and divide – and the challenges of experiential learning are a matter for rueful reflection. 'Contextualisation' is also an issue, but power and powerlessness (in relation to identity) are the strongest and constantly underlying theme.

The five-day seminar, which will be described in considerable detail (though not at such length as the workshop in Chapter 5), was organised by a London-based international NGO. It was designed for African women trainers who were already experienced practitioners in the field of community education and who wished to add to their training repertoire the approaches, ideas and skills of conflict transformation. The following account, again based on my journal notes and written soon after the event, is a mix of narrative and reflection, summary and detail. It was written from my viewpoint – it could not be otherwise – but I tried to do it honestly. I realise that by making it public I am exposing myself to all kinds of criticism, but I think that the issues it raises are sufficiently important to make that risk worthwhile. I certainly learned a great deal from the experience.

My account begins with the background to the workshop and contains a good deal of commentary, both within the narrative and after.

Before the Workshop

When I was originally invited to be one of two facilitators in this training for trainers, I welcomed the opportunity to increase my small experience of working in Africa, but was quite clear that my participation would make sense only if I worked with someone who knew Africa intimately and would be accepted by the participants as doing so – in other words, an African. I also wanted to respect sensitivities about racism and colonialism. I knew, and liked and respected enormously, a Ghanaian woman trainer, Cleo. We had often said we would like to work together. I contacted her and she was excited by the prospect of this training. We agreed to do it. She was in the US at the time, but we did some planning by phone and correspondence, had a brief meeting when she was in London, and planned to spend two days together immediately before the workshop.

Two days before this scheduled planning meeting, I received a call from Cleo to say that someone very close to her, whom she had regarded as a second father, had died unexpectedly, that she was devastated, and that his funeral would take place during the week of the workshop. It was clear that she could not come to Zimbabwe. It was also clear that it was too late for the workshop to be cancelled (since all the tickets had been bought, and some of the participants had already begun their journey), and far too late for me to find another co-facilitator, or for the organisers to make other arrangements. This left me, as I felt it, with no responsible choice but to make the best of a situation which I would have avoided at all costs – undertaking a week's facilitation alone in a training for trainers (always a daunting task), a single European trainer with a group of African women.

I decided that the best thing I could do, both for myself and for the participants, was to explain the position to them and to ask for one person each day to help me adjust or remake the next day's agenda (in the light of feedback from base groups) and co-facilitate the following day. Clearly this arrangement would be less than ideal, since the normal overall co-planning would not be able to take place or full co-responsibility be assumed. On the other hand, it would be a way of utilising and acknowledging the expertise contained within the group, and sharing the facilitator's load and power, at least to some extent, and modelling a co-operative way of working. I felt I needed ongoing input into the agenda from an African perspective – and I needed not to be in the uncomfortable and potentially offensive position of a lone European trainer in an African group. I did not consider such a position as appropriate and did not wish to be seen as considering it to be so. What I did want was to respect the participants and be respected in return.

I arrived at the conference centre outside Harare early on the Saturday morning. We were due to begin the following evening. The venue had been chosen by local partners who had considered it suitable. My first impression of the site was of its pleasant homeliness: attractive gardens, pleasant places to sit and work outdoors, friendly dining room, what seemed adequate bedrooms and washing arrangements, and very comfortable beds. I realised that, as is often the case at such events in Europe, some participants and some staff would be sharing rooms. Given the strain I knew I would be under, I was glad that I would not. I learned that the telephone lines were poor and that transport into the city was unreliable. These disadvantages seemed to me regrettable but manageable. What seemed more serious was that the plenary room was rather small for our purposes, and hot, and that the interpretation equipment necessitated the use of table microphones and thus an undesirable degree of formality and crowding in the seating arrangements. I felt some despair and irritation, then decided there was nothing to be done but to make the best of it, arranging the tables in a horseshoe shape, bringing in as many fans as possible and resolving to go outside for games and group work.

Composition of the Group; Identities and Roles

The 25 participants came from Burundi, Cameroon, Madagascar, Gambia, Kenya, Liberia, Mali, Ruanda, Senegal, Sierra Leone, Somalia, Sudan, Tanzania, Togo and Uganda. The absence of anyone from Zimbabwe itself was the subject of adverse comment from participants and greatly regretted by the organisers, who had relied on their local partner organisation to make the opportunity known to appropriate groups and individuals.

Maintaining a sense of equality in the workshop between French and English was going to be important, especially showing due respect to French-speaking participants who, as things had worked out, were in the minority. Two interpreters, recommended by the local partner organisation, had been hired, and although I facilitated plenary sessions in English (as the strain of trying to do it all in French would have been too great), I was able to understand French without interpretation and to use my French with individuals and small groups. I also made sure that everything went on flipcharts in both languages and asked participants to tell me if any of their language needs were not being met.

Since African women usually divided by language were given the opportunity of working together in this seminar, and since Africa is no doubt blessed with as much cultural variety as any other continent, this was for all of us a 'cross-cultural' event. However, the biggest cultural difference – and difference in perspective – was the one between me and my European organisational colleagues on the one hand and the participants on the other. Two African staff members from the organisation, Mbiya and Faith, who were responsible for different aspects of the organisation's Africa programme, were there, at their own request, as participants, and played a very helpful role in that capacity. However, it seems, retrospectively, that however useful to them and well explained their presence and role were, it must have reinforced the idea of Africans being excluded from the leadership. (I had not been very comfortable about their coming in that capacity, because of the potential for confusion about roles, but had agreed because they thought the experience would be helpful to them.)

Opening Session and Day One

The first 24 hours of the workshop were a struggle for me. In spite of all requests to the contrary, many participants came late, so that introductions and explanations had to be carried over from the first, already curtailed, evening to the following morning, and were not as full as had been intended. After a clear explanation of my colleague's absence, a heartfelt expression of my regret – on their account and mine – and a request for the group's understanding and support, I began with an explanation of the workshop's purpose, assumptions and agenda, and moved on to some consideration of group dynamics and of attitudes to conflict. This was followed by some work on communication skills, which I explained as constituting the building blocks for conflict resolution. In the plenary evaluation at the end of the first full day, participants expressed satisfaction with the day's content, the participatory methods used, the contributions of fellow participants, and the 'useful ideas and techniques', but they complained that the learning objectives of the workshop had not been made sufficiently clear and that they lacked a theoretical framework. In addition, they wanted greater clarity on the difference between training and training for trainers, and they wanted to move more and work more in small groups.

Here I feel the need to comment on my use of the word 'they'. In group evaluations, without a show of hands on each point made, it is not always easy to tell how many unheard voices would be in support of those which are heard. Still, the opinions expressed and described above had some backing and were supported by the feedback from the base groups which met at the end of the plenary session. Here there was appreciation for my facilitation style. I was seen as patient, articulate, sensitive to participants' needs and 'not wanting to impose'. But the sense of a lack of theory and definitions was reiterated, along with the need for greater clarity about the aims of the workshop. In addition, there was an expressed wish for 'a gender dimension', a focus on the role of African women. I noticed two responses in myself. One was of puzzlement. I had given advance explanations, which to me had seemed quite careful, even laboured, about the seminar's concept and aims, my approach to theory and terminology, the fact that for the rest of the week we would be working mostly in small groups and that the focus of our work (for instance, gender issues) would come from participants' own choice and experience – and yet my explanations had apparently failed to communicate these things. My other response was a determination to meet as fully as I could the group's clearly expressed needs – and to be seen to do so.

Day Two

Having conferred with the organisers, Jen and Kirsty, and enlisted their help in reproducing materials, I was able the following morning to present participants with a substantial theoretical tract, a list of workshop objectives (which was simply a re-presentation of the outline agenda) and an extremely detailed week plan, intended for a manual. I also repeated my initial assurances that, from the second day on, much of the work would be done in small groups. I gave a fresh explanation of the way this 'training for trainers' had been conceived and how gender and other issues could become the focus of our work. I explained my view that definitions and terminology, though covered to some extent in the theoretical tract they had been given, should not pre-empt our own discussions and conclusions, and that the one word we had discussed so far, 'conflict', was a word in common use, not to be imprisoned in some narrow definition; that what constituted a conflict was very much a matter of individual and cultural perception, or social construction, and that what we were mainly concerned with was finding ways to avoid the violence and destruction which all too often accompany conflict. To offer any other definition would for me have been to be sucked into something which I considered counterproductive. I explained that for the same reasons I had not wished to begin with a theoretical lecture, that for me theory was more usefully constructed in dialogue and on the basis of experience, and that to begin by lecturing would have been to model the kind of relationship between facilitator and participants which I did not want.

All this was very uncomfortable. It seemed to be my educational understanding and professional integrity versus the participants' demands. It also felt as if I was receiving double messages, one about the importance of participatory methods and group work – elicitive processes – and the other about the need to be told things, to be offered something ready made. Of course these two wants are not necessarily contradictory, and they perhaps represent, respectively, modern and more traditional approaches to learning in Africa – as elsewhere. (I remember similar debates in the North Caucasus, and from my work in England.) Perhaps I need to be less unbending and give people the reassurance – and perhaps clarity – which an opening lecture could provide, but something inside me says that to do so would be to give the whole workshop the wrong frame.

I could, I think, have used my 'power and conflict resolution' model to present, in an interactive way, the different elements of the week's agenda in a visual and at the same time theoretical way. Maybe that would have satisfied, to some extent at least, all these conflicting wants. At the same time, I wonder to what degree these first-day complaints were just that: the vehicle for a kind of early manoeuvring in power relations, which are always particularly sharply felt in training sessions for trainers and which had taken on an added edge because of the North–South dynamic which emerged during the course of the week. With hindsight, I think it was this dynamic, these perceptions about power, that came like an invisible wall between us, distorting what was heard and creating in some a resistance to what was being offered.

I had explained my idea of inviting participants, in Cleo's absence, to help with the planning and facilitation, and had asked one of the base groups if they could suggest one of their members for the following day. Their message to me after their meeting was that they were not willing to go along with this plan, since the relationship could not be an equal one as they had not been party to the planning of the workshop. At the time I fully accepted this response, recognising the truth of what was being said, while very much regretting that the aims of my suggestion would not therefore be accomplished. I was anxious, too, at the prospect of having to carry the full strain of facilitation for the entire workshop. What I was more concerned about, however, when I thought about it later, was one participant's suggestion that my proposal had been some kind of 'power play'. Presumably she meant that I was insincere in my proposal, wanting simply to appear to share power while not really meaning to do so. I think now that I should have challenged this suggestion. At the time I scarcely registered it and felt I had no choice but to express regret, accept what was being said as representing the considered response of participants and carry on alone. I did, at the end of the day, re-explain my proposal and invite any group or individual who felt able to help me to do so, but received no response.

There had been some complaints and suggestions from the base groups about practical arrangements: complaints about the telephones and food, suggestions for starting sessions at different times, and a request that *per diems* (daily allowances) should be paid, in spite of the acknowledged fact that it had been made clear in advance that they would not. Kirsty promised to speak to the kitchen about the food and Jen offered to send telephone messages, but no reference was made to the matter of *per diems*. In the organisers' view it had been clear all along that there would be none, and apparently they thought that nothing was to be gained by a discussion. In hindsight this could be seen as a mistake.

At the end of this second full day the plenary evaluation was extremely positive. The process had been lively, the work interesting, and time had gone quickly. The group work had been enjoyable and productive – especially the role-plays – and it had been good to be outside. Both I, as facilitator, and Jen and Kirsty as organisers, had been seen to respond to participants' needs. The objectives of the workshop had been made much clearer, the 'scientific quality' of the day was appreciated, and it was felt we were working on a 'usable module'. One person repeated the request for gender issues to be taken up more concertedly, but no one spoke in support of this request, and when later I invited her to look at the agenda and come back to me with a proposal, she in fact came back with the view that the agenda provided plenty of scope as it stood.

We seemed to have come to the end of our plenary evaluation and I was suggesting that the base groups could now meet, when one woman declared a need to raise for all participants the question of *per diems*. I reminded her of our 'speak for yourself' ground rule, but she insisted that there had been much out-of-sessions talk on the matter and that she really could speak for the group. They all felt that it was a bad policy not to offer *per diems* – a failure to recognise participants' needs – and that the organisers needed to hear that. It was a question of justice. One of the base groups had already raised the matter, yet no response had been made. I invited my colleagues to respond now. Kirsty, by temperament exceptionally open and friendly, described in what seemed to me a straightforward way the organisation's policy on *per diems*. Referring to the communication that participants had received on the matter, she concluded that they had therefore apparently chosen to participate on that basis, implying that to her there was really nothing more that could usefully be said.

This response was greeted by an uproar and a catalogue of complaints about the venue, accommodation, facilities and food, and what was seen as a general disregard of what participants were entitled to expect. It was suggested that this lack of respect was due to the fact that they were Africans and women. I continued in my facilitator role, since these were not matters for which I was responsible, but I felt the strain of distancing myself from my colleagues and my own opinions, and my feeling that Kirsty was being very unfairly treated, in view of the immense care she had taken over all the arrangements, both generally and for individuals, for whom nothing was too much trouble. I allowed myself to offer one piece of information, which was that the venues used for the European training workshops which I had facilitated for the same organisation had been considerably less pleasant and comfortable than this one.

This statement seemed to have no impact – maybe it was not believed – and eventually Jen was asked to speak. She reiterated the organisation's policy on *per diems*, but chose also to accept some responsibility for its application in this circumstance, as well as for other arrangements. She said that the cost of the seminar was high – and gave the figure – and that a choice had been made to spend a major amount on travel, bringing together a group of wide geographical scope, to make it a real pan-African event. The venue had been chosen with the help of the organisation's local partners. It was a venue which they had felt to be pleasant and comfortably adequate, and they had assured Kirsty that it was a venue frequently used by comparable groups. (One of the difficulties here was that the partner organisation had signally failed to deliver the kind of reliable support and advice for which they had been contracted and paid, and that Jen and Kirsty felt unable to announce this fact.)

Jen's words in turn caused even greater indignation with some participants. The mention of the cost of travel seemed to suggest to at least one of them that Africans were being blamed for the appalling inadequacies and costs of African air services, which she saw as yet another colonial legacy. Eventually, however, emotions subsided. I said I was sure that the feelings of participants in this seminar, about *per diems* and other things, would be noted by the organisation, and borne in mind in future policy discussions – which Jen confirmed – and this assurance was greeted with satisfaction. Participants declared themselves pleased to have had this discussion, glad to have aired their concerns. Several of them afterwards said it had been hard on me to have to hold the process. In fact, I remember I had at one point asked for an adjournment because I felt too tired to go on, but my request had been ignored!

The way that individual participants related to me outside sessions was in marked contrast to this apparent unresponsiveness. Many stopped me to encourage or congratulate me, a few to ask me how I was coping or to sympathise with me over the heavy load I was carrying. One even told me she had woken in the night and wondered how on earth I was surviving! At meal times everyone was very friendly – to Jen and Kirsty too.

Days Three and Four

During our second day's agenda, focused on problem-solving, the twin questions of power and justice had been raised. I had acknowledged their key importance and pointed to the fact that they would be our focus for most of the second half of the week – an assurance which was positively received. When, on the Wednesday and Thursday mornings, I presented the Goss-Mayr diagrams, they were clearly of great interest to the group, and yet I felt resistance, particularly when we discussed the option for violence or nonviolence. On thinking carefully later, I realised that this resistance had come mostly from one person, but it felt to me as if it created, or maybe represented, a *dynamic* of resistance. It was clear from the way participants worked with the models later that they were in fact relevant and useful, but in the plenary sessions in which they were discussed I had the same feeling that I had had earlier – that something was obstructing or distorting my words' reception. For instance, when I suggested that international solidarity could help in a campaign, one participant took that as encouraging dependency. When I responded that the choice was with the local organisers she remained dissatisfied. However, when other participants joined in the discussion, it gave us the opportunity to clarify the principle that the leadership and agenda should stay with those who initiate a campaign.

In spite of the difficulties of these discussions, the evaluations of the Wednesday and Thursday were reassuring. According to these, the models had been new and helpful and the content of the discussions had been good. The participants had also appreciated the chance to stay in the same working groups (I had expressed some anxieties about that, but accepted what seemed to be a strong majority view) and the work had been engaging and productive. The role-plays (which they had done alone in their groups, rather than as performances for each other) had been powerful and had brought new understanding at an emotional level. I was thanked for the strength and quality of my facilitation and was considered to have been articulate and confident. In addition, the food was considered to have improved and the Wednesday afternoon's trip to the city had been enjoyed. (Here Jen and Kirsty had had an opportunity to demonstrate their care and responsiveness. One of our party had had her purse stolen, and returned very distressed. Kirsty and Jen had decided right away to replace the stolen money, and this was greatly appreciated by the person concerned, and indeed by the whole group.)

That night I dreamt I was preparing for the final party and that when I looked at myself in the mirror I saw that I was, after all, African. In my dream I was overcome with relief. Waking was hard.

Day Five

[My account of this last day is more reflection than narrative.]

That morning I recounted my dream of the previous night to one or two of the participants, wanting somehow to communicate what I felt. One of them came to me later and made me a gift of one of her own very beautiful African dresses. I was deeply touched.

On this fifth day, our plan was to spend the morning on group work on role-playing nonviolent action which would be appropriate for the cases the participants had been working on. The afternoon was to be used for thinking about what was necessary for recovery after violent conflict. It emerged, however, that several women wanted to leave early and none wanted to miss anything, so in the event the processes intended for the Saturday morning – final evaluation and closure – had to be squeezed into the Friday afternoon, with detrimental effects.

I have since reflected a great deal on the dynamics within the group, which made it so difficult. The whole question of individual opinions and 'group opinions' is a vexed one. I referred to it earlier in relation to evaluation. By the end of the week I was of the opinion that one particular person in the group played a key role in defining the apparent relationship of the group with me. As I have gone through my notes and recalled who said certain significant and 'impactful' things, I realise it was this same person. By the position she took she drew behind her, at certain times, a few other powerful participants, but I think those others, without her setting the pace, might have responded quite otherwise – and to an extent did. Yet this one strong thread of resistance and subtle attack had a major effect. How do the quiet and the silent voices relate to the loud voices? How can they be detected and valued? The base group feedback process was designed to give them a chance, and I am sure most women spoke up in the base groups, but were all voices represented in the base group reports? I guess that was a question of who was facilitating. There were a few women who stood out as different in style. One of them would make separate, personal evaluatory speeches of an enthusiastic kind which tended to be laughed at by the others; another just seemed to be separate, engaging with the others, yet in some way remaining contained within herself and not following the emotional movements of the group.

There is such a thing as a group dynamic, but no such thing as a group mind, except in the sense of an acknowledged consensus which has been worked for and which somehow expresses or takes into account the different minds of individuals. Such consensus can be more or less easy to achieve. When it is difficult, this can be on account of one of two things, or a mixture of both: a general inability or unwillingness to make a particular choice (wanting to have the penny and the bun) or a polarisation between individual choices within the group. Thus when we had to decide whether to end early, on account of the early departure of a substantial number of participants (because they had chosen to opt for more convenient flight schedules) some insisted that we should do the impossible, curtailing nothing and including everyone in everything, and complained at every choice or formula for compromise; others took a strong position one way or the other. Who, in this circumstance, was responsible for finding a way forward? In my theory, that responsibility was shared (and here responsibility is the other side of respect). In practice, I kept trying different formulae and combined my final best offer with a process reflection, pointing out that it was not possible to meet every conflicting want, and that I could not make it so.

The option that was eventually chosen was that we should all end early, with the result that our work on 'recovery and healing', which was of great importance in the group, had to be curtailed, and some participants were left dissatisfied and with some raw emotions. I felt sorry for this, though only partly responsible. It was a consequence not only of my compromise proposal, but also, more fundamentally, of other people's choices over flights, or insistence that we should all be together for a full evaluation, which then necessitated the shortening of the other afternoon work. No one, I now reflect, offered to share the responsibility for this lack of time for the question of healing.

(The inadequacy of our session on recovery and healing confirmed for me the subject's importance and complexity and confirmed the need for more in-depth training work on specific aspects of peace-making. Recovery and healing would be one such aspect.)

One might expect that trainers in training would more readily accept co-responsibility for the agenda and group process but, in my experience, trainers who are having the chance to be in a different role often like to behave as if they are on holiday, as well as competing with each other in knowing better than the trainer(s) for the workshop. In addition, this group was composed largely of feminists, advocates and trainers who specialised in assertiveness. Maybe there was something of a contest as to who could be most assertive, with me as a kind of substitute male authority figure! I need to bear these possibilities in mind when I try to estimate how much of the dynamic within this group of participants, and between them and me, was or was not a question of culture, historic relationships or race, or how much it related to aspects of my behaviour. Maybe the African–European relationship was to some extent a focus or cover for some more universal class–teacher dynamic. (I am acutely aware of how far all this is a matter of speculation – and maybe projection – on my part.)

I also find it interesting to speculate whether antagonism towards the European organisation and leadership of the workshop played a role in unifying potentially conflicting elements within the group – for instance, the French- and English-speaking sub-groups. In the final evaluation, one participant (the usual one), complaining about the standard of translation and describing what she saw as the lack of care taken over it as an insult to Africans, laid the blame for linguistic barriers in Africa at the door of colonialism.

Evaluation

Our final evaluation was done both in base groups and in plenary. The casework of the final morning, with action plans and role-plays, was very positively evaluated. Assessments of the whole week were appreciative of the workshop's overall style – my 'allowing problems to be aired' and responsiveness to participants' needs, my 'knowledgeable and articulate' facilitation, my approach and methodology, in particular the participatory processes used – the practical usefulness of the knowledge generated 'for day-to-day conflicts' through the opportunity in group work to look at real conflicts from the participants' own experience. There had been 'much learning of fundamentals'.

In spite of the sense of solid and useful learning, of new and practical knowledge, there were still complaints about a lack of theoretical grounding. One group saw this as a question of time, suggesting that the balance had been right and that different aspects of the work, including theoretical, would need to be developed in follow-up workshops. This made sense to me, and I want to consider seriously the idea that a more theoretically structured opening would be helpful. One group said that the issue of power and justice had not been adequately handled. Since the daily evaluations on this part of the workshop content had been entirely and enthusiastically positive, this came as a surprise. Maybe the subject was of such importance that the time available was experienced as insufficient (but this feeling was not expressed in the daily feedback). Perhaps our brief exploration of the choice between violence and nonviolence had provoked feelings which were not dealt with. From the subsequent remarks of Mbiya (one of the two African staff-participants) and associated memories from our discussion during the workshop, I see reason to favour this latter explanation. My brief presentation on nonviolence as an approach was intended to be descriptive rather than prescriptive, but I also made it clear that our purpose in this workshop was to examine and develop nonviolent rather than violent strategies and methods, which was of course well known but may still have been felt as some kind of put-down of violent struggle and therefore of struggle itself. (This is one of those 'swimming in treacle' areas where all the explanations in the world seem to communicate little when people from the old colonial nations are talking with those who have been colonised and are still suffering the consequences.)

Another area of complaint was that I had not adequately 'contextualised' my material. I had tried repeatedly to explain that I, not being African, was deliberately choosing non-African examples, (though certainly not mostly European) to illustrate what I was presenting on the understanding that the participants were the Africa experts and it was for them to bring in their own experiences and examples, and to test and apply the models and theories under discussion in relation to their own examples. For me this was very much an issue of respect – to know the limits of my own expertise and competence and to acknowledge theirs, as I explained. I considered it would be of interest to them to hear, for instance, of struggles for justice in other parts of the world, illustrating the universal dimension of some issues and experiences. Instead, it was taken, by some at least, as a measure of my lack of interest in African experience – or maybe simply of my ignorance. (Earlier in the week, the 'iceberg' diagram (Figure 4.4) I had used to delineate the processes needed for problem-solving had been felt by one group to be inappropriate for Africa. I had, in fact, wondered about this, but had reasoned with myself that I probably live no closer to icebergs than the participants do.)

The final criticism about the workshop content, which came from one group out of four, was that I had failed to respond to repeated requests for a gender focus. Here I realised – and explained – that I should perhaps have told the whole group about my conversation with the one woman who had repeated the request for a stronger gender focus and later withdrawn it on the grounds that the framework already existed for it. I had not done so because I had thought she spoke only for herself. In practice, I think the women were indeed free to work as much as they liked on gender issues, and did so when they chose.

The evaluation of practical and organisational arrangements surrounding the workshop was much as could be expected, though communicated with less anger than before. Criticism of the translation provision was new. I felt for the two interpreters who were obliged to translate the view that they needed to improve their fluency! Much of the feedback given during the week seemed, from my cultural perspective, quite harsh in its frankness. It had the advantage of clarity and the drawback of being bruising, though the tendency to be bruised may also have been cultural. I am acutely aware of my inability to weigh these things.

After the Workshop

Although the workshop had clearly been useful to the participants, who had brought to it a wealth of experience and skill, it had major emotional and ideological repercussions for those of us who returned to Britain. Being somewhat obsessive about questions of justice, I found it difficult to be associated, as it felt, by geography and race, with colonialism and oppression, and experienced this as persecution. Whereas the group clearly experienced me as powerful on account of my role and Europeanness, I was more aware of their power and my isolation, both functional and racial/cultural. It felt to me that their power was not only numerical but moral. It was just such a polarisation around power, victimhood and guilt that I had aimed to avoid, or at least minimise, by working with Cleo, by using base groups to devolve power, and by working with a largely elicitive process.

I had been unhappy with myself that, although I had given as one of the workshop assumptions the notion that at times our own process should be the focus or our learning, I had not fully lived up to that promise. I had, it is true, allowed the anger of participants to be voiced in plenary and had provided channels for ongoing feedback through the base group and plenary evaluations. However, I had not named or confronted what I saw as the lack of care and respect which I felt in some of the attacks made on Kirsty and Jen, or invited an alternative explanation. I did make some reference to these things, the morning after the turbulent plenary, noting to the group that the ways in which we expressed differences amongst us provided us with material for learning, but I did not, as the week progressed, make explicit my interpretations of the underlying dynamics within the group, or raise them for discussion. In the case of the attack on Jen and Kirsty, I felt that to make my judgements known would conflict with my function as mediator/facilitator, and in terms of the more general dynamics I had judged it as beyond my power to raise the matter in such a way that my interpretation would be understood or taken seriously, or would not be felt as the final affront and demonstration of incomprehension. To be constructive, any approach to conflict (or education) needs to be based on a realistic assessment of possibilities and likely outcomes. In this case it seemed to me that I had to make a choice between completing a clearly useful workshop agenda and running the risk that things would fall into chaos and recrimination.

In a subsequent meeting with Cleo I shared this concern with her. She supported my judgement at the time and dismissed my doubts. In her view I, as a single European trainer, could not usefully have confronted what was happening. Maybe the two of us could have, had she been there, though then the dynamic would very likely have been quite different. I still wonder, though. One problem is that it takes distance for things to become clearer – both in terms of what was actually happening and what responses could have been made. For instance, had I been clearer at the time about the role that seemed later to have been played by one particular participant, I could have asked her privately what was going on between us. This could have had an impact on the whole dynamic. (It would not, however, have provided an opportunity for the whole group to reflect and learn.) I could also have tried harder to get advice from the organisation's African staff members who were participating, though they seemed to want to avoid such conversations.

I have thought much about the criticism of the perceived 'lack of contextualisation'. In the first place I could not have done otherwise than I did, in asking the participants to speak and work from their own experience. I am not an Africa expert and had expected Cleo to fill that role, if necessary. Secondly, I still find my logic sound: that the contextualisation was best done by the participants; and this was proved to work well in practice. Yet I cannot ignore the objecting voices. They came from a wonderful group of women, deeply committed to ideals that I share. Maybe they needed me to give African examples not in order to build conceptual bridges, but rather to prove myself and to build a bridge between me and them. Perhaps, seeing their continent as having been subjected to oppression and belittlement, they wanted clear signs of recognition, in particular, the recognition of the importance of their own situation and experience, through the use of African examples. Maybe I caused unwitting offence by seeming not to think African experience sufficiently significant to be cited. (I did talk about South Africa, which is, however, viewed with some suspicion by other African countries as being Westernised and the favourite of the West.) Whereas most Europeans are eager for examples from other continents and cultures, I have experienced resistance to this (universalising?) approach also in the Middle East and in the former Soviet Union. My tentative conclusion is that this resistance may be expected when participants feel their own dignity and identity to be marginalised.

Jen, Kirsty, Mbiya, Faith and I had a further evaluation meeting once we were all back in England. Mbiya and Faith had remained silent during the plenary evaluations in Zimbabwe. During the workshop generally, I had found them very supportive, Mbiya volunteering to do much of the French flipchart writing for us, and Faith playing a constructive and at times bridge-building role in plenary discussions. When it came to this meeting in London, however, whereas Jen had hoped they might offer some sort of intermediary perspective – organisational at the same time as African – in fact they stood firmly behind the viewpoints expressed by the more vocal of the other participants. Mbiya noted the sensitivity of non-violence as a subject 'which cannot be neutral'. In relation to contextualisation, Faith proposed that in future participants could be invited to write pieces about their own conflict experience, for circulation in advance. She and Mbiya also suggested having a trainer in reserve for future workshops, in case one had to withdraw. Otherwise they concurred with the points made in the end-of-workshop evaluation. When Kirsty expressed her puzzlement at the strength of the anger directed at her and Jen about the facilities and lack of *per diems*, Faith and Mbiya were clear in their view that the two organisers had acted insensitively and provocatively, and emphasised the negative impact of the inconveniences and discomforts of the venue. Overall, it seemed that, away from their African co-participants, they wished to stress their primary identification with them rather than with their colleagues. It was clear that the tensions felt during the workshop were reflected within the organisation itself.

I later received a digest of the participants' written evaluations. It was interesting to note that the negative aspects of the individual evaluations were relatively slight in tone and proportion. I was left in no doubt about the usefulness of the workshop, which several participants wrote later to confirm, listing the ways they had already used the experience and proposed to do so. The materials used in the workshop have been re-used by participants working in Africa, who felt that the workshop had prepared them to become trainers in this field. At a conference a year later, I was delighted to find that one of the speakers was the Somali participant, who described how, in her work with women's groups, she began by getting them to analyse the injustice they suffered, using a diagram with an inverted pyramid supported by pillars. The Goss-Mayr models had clearly travelled well.

CONCLUDING REFLECTIONS

This workshop confirmed me in my belief that the 'cultural barrier' between Western trainers and groups from other parts of the world is less about the substance of what they can offer than about perceived power relations and associated attitudes. There may be real cultural differences which make communication more difficult than usual, but they are not, from my observation, insurmountable. The trainer's ignorance may be irritating, but if s/he admits it and listens attentively, it is unlikely to prove fatal. Resistance to an unexpected pedagogical approach will be overcome if the approach is seen to work. The real problem lies in interpersonal perceptions and relationships, which are at the same time more than personal, carrying, inevitably, an enormous amount of historical, political and economic baggage. The barrier is created by destructive power relations which, though rooted in the past, are perpetuated now when workshops are funded and organised from the West (of which the *per diem* issue was symbolic). The choosing of Western trainers is, not unreasonably, seen as the exercise of that unequal power. These tensions will disappear only when the wider, underlying conflict created by geopolitical power asymmetries has been transformed at its root, so that the exchange of trainers is two-way, and when the wounds of the past have been healed.

It is on such experience that the views expressed in Chapter 3 are based. If I consider how I would feel if I participated in a women's workshop and found that the only facilitator was a man, I can understand how inappropriate it must have felt for African women to have a single, European facilitator. In those circumstances, even to have a mixed facilitation team would seem questionable. So I am left with a very large question mark over the idea of Western trainers working outside Europe and North America. For myself, I need clear reasons and a lot of persuasion. I do it infrequently and never without a colleague from the region in question. (I should note here, however, that when the facilitation task is one of mediation, rather than training, coming from outside the region in question is likely to be an advantage – though coming from the West may still be less than ideal.)

Although the loss of a co-facilitator had particularly painful consequences in this case, the more general lesson remains: it is important for the power and responsibility of this job to be shared.

The question of 'women's issues' and how to approach the framing and content of all-women workshops is one which has recurred in my work and, although it has exercised me, my position has not changed fundamentally. I consider it of great importance that women should have time to work together on the issues that most concern them, but I do not see that it is appropriate for them (including me) to be told what those issues should be. Not long ago I was persuaded to co-facilitate a workshop for women living in India and from neighbouring countries. Halfway through, one woman asked why we had so far focused all our attention on 'women's issues'. My response (as it had been during the workshop in Zimbabwe, when the question was about why we had not done so) was to remind her that she could introduce whatever issues she liked. To restrict women's choice because they are women would seem to be perverse and counterproductive, if the goal is to emancipate women.

Whatever the issues women choose to address, this experience of working with African women leaders and trainers deepened my conviction of the vital role women can and do play in transforming societies, and of the outrage of their abuse and exclusion. They made me wonder whether I should not devote the rest of my life exclusively to the emancipation of women and the promotion of 'feminine' values and ways of being. Although I have not done that, the passion that was fuelled at this workshop has remained, and is expressed in many ways in the work that I do.

7 Women Peace-makers: Transforming Relationships, Finding Power

Though the topics they address may be similar, training workshops take on a different aspect when they bring together participants from 'different sides' in an incipient conflict or, as in the case of the workshop described in this chapter, a war. Cultural differences may or may not be important, but differences of viewpoint and tensions between participants are bound to be considerable. Conflicts within the workshop group are likely to have particular significance and if the external conflict is acute and violent its trauma will be felt within the workshop. This chapter will illustrate and explore, through the account of a workshop for participants from what had been Yugoslavia, the challenges involved for participants and facilitators. It will also demonstrate the transformation of relationships and attitudes that can be achieved. In showing the mediatory character of the trainer's role in such circumstances and the intense and immediate significance of 'experiential learning', it will surface some ethical/professional issues for facilitators in relation to risk and responsibility. The inspirational and restorative role of workshops will be clearly illustrated and an indication of the impact of the workshop on the future lives and activities of some participants will suggest its importance to them.

The workshop in question was, like the one in Harare, a women's workshop. I was then, as I am now, working a good deal with groups from the former Yugoslavia, and in all of them women greatly predominated. Many men were directly involved in the war and a high proportion of the people who were active against it were women psychologists, teachers and journalists. In this account I discuss the difference it can make to work in a women-only group. Some of the participants were already active in the human rights and peace movement, but many were new to the idea of involvement in this kind of work, and new to this kind of workshop. They were all living in very difficult circumstances and were carrying much fear and grief.

The organising group of Scandinavian women (five in all) participated in plenary sessions of the workshop, and played a supportive and servicing role. I allowed myself to be persuaded to facilitate this workshop on my own, with their backing. My misgivings about this were confirmed and I came to see my agreement as unwise. There was no problem for me in being accepted by participants, but the moral power of victimhood, which from my perspective seemed to be operating in Harare, was experienced in another way.

We were housed in a guesthouse beside Lake Balaton in Hungary. In these delightful surroundings, the women had a chance to let go of some of the misery and tension they had carried for so long; a time away from the necessity to cope. It was an immensely charged and cathartic week, though it also involved much hard thinking and planning.

One major challenge was to deal with the conflict present in the group. I had been aware of the potentially explosive mix of participants, but somehow had not really registered just how much we were all taking on. My account of the workshop, again written soon after the event with the help of detailed journal notes, reflects the emotions of the week. It is relatively short, describing two key episodes in detail, and discussing other aspects of the workshop in more general terms. Its emotional tone reflects the mood of the workshop, and the 'spiritual' aspect of these events is once more a matter for reflection. As elsewhere, the names of people have been changed.

Apart from the four Swedish, one Finnish and one Hungarian woman representing the European organising group, and the workshop leader from England (me), the group that gradually came together at a conference centre near the shores of Lake Balaton was made up of women from different parts of what was once Yugoslavia. There were five Hungarians from Vojvodina and one from Croatia, one Serb and one Albanian from Prishtina, one Croat and one Serb from Belgrade, and Croats and Bosnians from Zagreb and Osijek in Croatia and from the border town of Zupanja, which had suffered years of constant bombardment. Ages ranged from late teens and early 20s to middle age and beyond, which, together with the variety of styles, personalities and professions and the mix of deeply religious women – Protestant and Catholic (no Orthodox or Muslim) – with atheists and agnostics, made for a group of great complexity. Two things united us: we were all women and we all longed for an end to the misery in what was once Yugoslavia.

Although I had been well aware of the intention to bring together a regionally and nationally mixed group, and to draw in women who had not yet experienced similar workshops, I had not fully internalised what that would mean in terms of fear, suspicion, pain and hostility. As part of the introductory process on the first evening, participants spoke – first in twos and threes, and then in the whole group – of their hopes and fears for the week. What they hoped for was trust, openness and tolerance; what they feared was mistrust, lack of openness, and conflict. At the word 'conflict', one young woman exclaimed that she was shocked that anyone should think conflict possible in such a group. This gave me the opportunity to say that I fully expected conflict, that it was my experience that groups started with a determination to be united and that in the event it was through conflict in various forms that they forged a stronger, deeper unity, that we had succeeded in bringing together a group which mirrored the mix of nationalities of former Yugoslavia and that therefore we were bound to experience some of the tensions of the current situation there, that we had come together to learn about creative responses to conflict and that learning by doing would be the most fundamental learning, and that I was confident we could do it – manage the conflict, respond creatively – and that I wanted participants to trust themselves and the process.

The women engaged with a will in the first full day's work, but I sensed their tiredness and stress. By the afternoon, rumour had reached me that the Hungarians from Vojvodina had told the Hungarian (from Hungary) organiser that they suspected the two young Serbs of being spies. Before I could arrange to meet with them to discuss their fears, the storm broke. In the evaluation at the end of the afternoon, Maria, one of the Vojvodina women, having in various small but important ways, as much for personal as for national reasons, found herself out of tune with the larger group, exploded with agitation and said it was not possible for her to work in a group where some members supported the expansionist ambitions of the Serbian government, and where she felt that to be open would expose her to great danger. Olga, one of the two young Serbs – also not the easiest personality in the group, giving a first impression of sulkiness or arrogance – responded that she had not expected to be attacked in such a group and that she felt threatened by such behaviour. This, as Maria later explained to me, was the last straw for her. How could a Serb feel threatened? It was the Hungarians who were being threatened (and worse) with the suppression of their language and culture, with the confiscation of their homes so that they could be given to Serb refugees from Krajina, ultimately threatened with 'ethnic cleansing'. She left the room in tears. The group sat dumbfounded and I acknowledged the distress and fear that the exchange had both reflected and engendered, repeating that such conflict was to be expected, that the feelings on either side were understandable and should be respected, that I had every confidence that they could be handled and that it was good for all of us that they had come out sooner rather than later. As I left the session, I spoke with Olga, who was clearly upset at Maria's reaction and said she had not intended to hurt or offend her, but explained her own feelings at being, as she felt, accused.

I went to Maria's room and found her in great distress, sobbing that she was not used to such 'psychological workshops', that she had been already near to breaking-point and did not know if she could stand the pain of this experience. I held her, comforted her, apologised for any part I had played in causing her such pain, listened to her explanation – the suffering and fears that had led to her outburst, her indignation that Olga should claim to feel threatened. I tried to explain to her, from my own understanding of what I had heard and from Olga's clarification, what Olga had meant. Maria asserted that she had not accused anyone of anything – had spoken only generally – which I challenged, as gently as I could, repeating what I had heard. I also said that I knew Olga already, and that she was well known in the peace movement (which Maria correctly said did nothing to prove she was not a spy) and that I was quite sure she was an opponent of Serb expansionism and nationalism. Maria could not see why she had not said as much, and I pointed out that she, with her background, would consider that self-evident. At last I asked Maria what she wanted to happen and how I could help. She replied, 'Oh I know what I must do. I'm a Christian. I have to be reconciled.' I wanted to ask her if she understood what she was saying, but decided that would be impertinent. She gave me permission to outline her feelings to Olga, and to arrange a meeting. Then she dried her eyes and came down to dinner. I marvelled at her courage.

After the meal I found Olga and tried to help her understand (without betraying any confidence) what lay behind Maria's outburst and subsequent walking out, explaining that she had no experience of opposition or peace movement Serbs and needed to hear Olga say what she (Olga) took for granted: that she opposed her government's policy and actions – and had indeed suffered for her views and actions against those policies. (She had been suspended from her job for six months.) Olga was only too willing to accept both the explanation and the need. She and I were looking for Maria when we came upon the residue of the base group to which, by good fortune, they both belonged. They (one Croat and two other Hungarians from Vojvodina) were discussing together what they could do to help their two missing members. I said that I thought their greatest need was to feel accepted, to know that their behaviour was understood, and that they would still be welcome and would have the group's support in dealing with their conflict. Olga hung back, saying she did not want to interrupt their discussions, but they drew her in; and at that moment Maria arrived and joined the group and she and Olga began immediately to say to each other what they needed to, with the group's gentle support. Seeing the matter well in hand I withdrew.

By the end of the week these two women, the young Serb punk from Kosovo and the middle-aged Calvinist Hungarian from Vojvodina, had become each other's firmest admirers. 'I still find Maria's views difficult sometimes,' said Olga, 'but she's wonderful.' Olga's apparent sulkiness and distance evaporated as the week went by, and her engagement was thoughtful and committed (despite late nights and hangovers). Maria, through an astonishing piece of sustained virtuoso role-play, in which she took the part of a young single black mother of three unruly boys, became a star in the group (stardom consolidated by her impersonation of a noisy and prolifically egg-producing hen at our final party) and a major contributor.

I would like to have been more alert to the position of the two Serbs in our group. (We had not realised how the group from Serbia would be composed.) They felt very much isolated and besieged at times, as they told me later. They had no traumatised Serb refugee in the group to point to, either – no comparable suffering to place beside the suffering of some of their fellow participants. But they really handled their position very well and their very vulnerability in the group probably helped to disarm the prejudices of others. They also shared information, in a quiet way, about the pressures and difficulties of their own lives (which are very far from easy) which came as real news to some group members, and radically altered their perceptions of what it could mean to be a Serb.

The individual (and related) journeys through the week of Olga and Maria somehow symbolised the unfolding of the whole group process. When I announced at the beginning of the second full day that Olga and Maria had met, together with their group, and had reached a new understanding, the relief was palpable. I remarked that although it was unlikely we had seen the last of conflict, we could now feel the confidence that came from experiencing that we had the capacity to handle it. These feelings were reflected in the feedback from the base groups that evening. Trust was growing. Participants were far more confident and at ease, relaxing into the process.

Because of what had happened, I had decided, after some reflection and consultation, to change the order of the agenda, so that instead of proceeding with questions of violence, nonviolence and empowerment, we reverted to the subject of communication and the obstacles to it, including matters of identity, prejudice and strong emotions (followed by 'needs and fears mapping' – an exercise in empathy as well as analysis). One experienced participant, Vera, whom I consulted felt this was something we needed to do in order to address the question of individual and group identity and respon-sibility. Nonetheless, she was not convinced that we could handle it. Eventually we decided that it was a risk that had to be taken (the idea of risk-taking was a theme for the week) and in the event the day was all we could have hoped. Vera told me that evening that she had been very much afraid, but now felt completely satisfied with the outcome of what we had done – to the point that if there were nothing to follow 'it would be enough'. She said we had done what she had not thought possible and looked at these hard issues in ways that had not threatened participants but (in the words of another base group reporter) which had drawn them through a process in a way that did not hurt anyone but in which each felt she had been listened to. Since the questions we had worked through were also inescapably challenging, this was good to hear.

The crucial exercise, I think, had been one in which participants are invited to look at the question of identity and belonging (see Chapter 4). Each is asked to make a list of as many groups as she can think of to which she belongs or which help to form her identity, then to select the three which she considers most important and to write by each of the three something about it that makes her feel proud and something about it that makes her feel uncomfortable or ashamed. Participants are then asked to share these lists with one or two others. Individuals then share with the whole group anything they choose to, discussing, as appropriate, questions of difference, justice, and the need both for critical awareness and respect and sensitivity, in relation to their own cultures and those of others.

Later in the week participants chose to analyse together situations 'so hot', according to Vera, 'that we'd have needed the fire brigade if we'd mentioned them at the beginning'. There were differences, arguments, but, handled from a base of trust, they were no longer too threatening. Nonetheless, the women were surprised by – and justifiably proud of – their own courage.

This profound learning and changing, brought about by the experience of confronting conflict and building community out of difference, was further deepened by the experience and recognition of common pain. On the morning of our next-to-last day, as we shared stories of nonviolent action, we were stopped in our tracks by an explosion. In all probability it came from the use of dynamite on a building site or in a quarry, but in our group its effect was devastating, and the initial panic was followed by half-relief, nagging doubts and bitter tears. Once the women were convinced that there was no danger to us there, they were overtaken by the fears and grief of years – and of their present reality. The stories came spilling out: of cratered gardens, rivers where no one dared swim, useless wrecks of houses, lost relatives, the daily risks of humdrum activities. After much crying, comforting, smoking, singing, more crying, more cigarettes and coffee, this courageous group went back to work. That is how they cope – by carrying on. And at our party that evening we laughed as none of us could remember laughing – real laughter, joy, total silliness, complete relaxation – and closed with more tears as we lit candles of hope and longing.

Could all this have happened in a mixed group, or a men's group? I think not. But is that my prejudice or lack of trust? Is it possible to generalise? There were clear reasons for the intensity of feelings here. But in a mixed or male group could they have been expressed and handled with such freedom? Could they have been allowed to flow, to reach completion – both the pain and the comfort, the tears and the laughter? Probably physical contact is important in all this, and in our group it was unrestrained.

In my reflections on a gathering of trainers and educators (almost all women) from former Yugoslavia held earlier in the year, I remarked on what seemed to me a preoccupation with the personal, psychological aspects of conflict and reluctance to spend time on analytical and strategic thinking and the more political aspects of conflict. Although the word 'political' was used only pejoratively at

Balaton, there was no reluctance to engage with questions of inter-group conflict and public policy. Indeed, the applicability of the (Goss-Mayr) models to such situations was a matter for great satis-faction, according to the feedback of all the base groups. The cases chosen by the different working groups during the day were highly appropriate to the models (or vice-versa!), and the analysis and consequent strategies they produced extremely cogent. The theor-etical input given to provide a context for the work of that day had also been received attentively. Indeed, I remarked in these women no lack of enthusiasm for theory, only a determination to find practical applications for it. Likewise, our discussion on the theoret-ical and philosophical basis for nonviolent action, and its different forms, was valued for its relevance to participants' own experience – because it gave a name and a thought frame to what they had been doing instinctively, so affirming their already courageous and im-aginative efforts and helping them to think about them more clearly. In addition, participants considered their new-found knowledge to be practical because they saw it as transferable to their groups at home (especially with the help of the manual which was promised and indeed produced).

Review of the Workshop as a Whole

The assumptions on which this workshop had been based had been explained on the first evening and were as follows:

- The spiritual, emotional and practical aspects of the group's deliberations and experiences will be woven together, since they are inextricably linked.
- The group will become a community of learning, using partici-patory methods, drawing on the experience and wisdom of each person, and working in an informal and relaxed atmosphere.
- Analysis and imagination are both important, and laughter and gravity are complementary.
- The agenda which has been prepared is intended as a framework for the development of understanding, skills, resources and commitment, and it can be changed as the workshop goes along.
- The group's own experience of working together will provide important material for learning, and, when it seems particu-larly relevant or necessary (for instance, if there is a conflict),

what is happening in the group may, for a while, become the focus of its work.

I believe we lived out all these assumptions. Our days began, for those who wanted it, with a time of meditation, to which many – those who had already identified themselves as religious – contributed. But the depth of communication and emotion that characterised the discussions that went on in and out of sessions, and the passionately held values expressed, could also be termed 'spiritual', and one of the cross-boundary exchanges that took place in the group was between religious and non-religious members discovering what made the other tick.

The mixture of seriousness and levity, intense discussion and crazy games, helpless tears and uncontrollable laughter, was also a hallmark of the week, and gave us a feeling of wholeness in spite of and because of the grief that was present. It was a mixture we needed for the health of each individual and for our healthy growth into a community.

The experience of discovering and sharing so much latent knowledge, both in plenary sessions and in work done in small groups, was a powerful and exciting one. One participant said to me at the end, 'How did you do it? When you asked us a question, you wrote down all our answers. How could you know we'd get it right?' I explained that I worked from the assumption that they did know most of what they needed already, and that I had made any additions or comments I had felt necessary. Another participant, giving feedback from her base group, remarked, 'It's interesting that the things we've been learning have been things we already knew but couldn't use because something in our thinking was stopping us.' Drawing out and helping give shape to existing knowledge is a large part of workshop education, but in this workshop the exchange of information was also of great importance. Within the divided region of former Yugoslavia, it is hard for communities to know what is actually happening elsewhere, and what others are experiencing.

This group's sharp engagement at the conceptual and analytical level was well matched by its imaginative and creative energy, which meant that reports from small groups, which can so often be tedious and lacking in impact, were in this case fascinating, being both clear and colourful in their presentation; and in the breaks as well as in sessions, women sewed and drew, read poems, created symbols and danced. The variety inherent in the methodology of the workshop, and its participatory nature, together with the experience of handling the dynamics within the group, served, according to participants, to maintain both interest and energy. ('We saved a lot of money here. Often at seminars we get bored and go to the shops instead.') Besides maximising learning by the sharing of knowledge, participation aided its digestion. 'It's easiest to remember things you've been involved with. Working on our own cases and doing role-plays, that's what made it all work' (base group feedback again).

The agenda and its daily structure certainly underwent changes, both major and minor, in response to what was happening in the group and to needs expressed, for example, for more rest at lunch time. Responsiveness to need was something we all aimed for, and represented the living out of the value of respect which provided the foundation for our working and living together. One of the base group reporters, expressing her group's appreciation for the fruit which was provided during breaks, noted that it was not only the fruit itself that was appreciated, but the care which its provision represented. Through this kind of practical respect, through the experience of mutual attentiveness, through the base group process which provided a place for all voices to be heard, channelled, and taken into account, participants felt the healing and strengthening power which come from recognition and acknowledgement.

I hope that respect, acknowledgement and encouragement were among the things I was able to offer as workshop leader. With this subject matter particularly, it seems essential to try, at least, to demonstrate as well as talk about it. This is true for the whole group, as well as for the leader. For me, it was a particular pleasure to have someone say to me at the end, 'You're a wonderful pedagogue [am I really a pedagogue?] – not just because of your methodology, most importantly because of your attitude.' And I can say the same for the group. They were wonderful participants, not just because they engaged with everything with so much zest and intelligence, but above all because they lived out the care and acknowledgement which lay at the heart of all our work and aspirations.

At the beginning of the week I had the feeling that the working agenda we were proposing was in itself a profound disrespect for the women's real (apparent) need for nothing more than a good holiday – and I think more than a few of them shared the same feeling. However, as the days went by, it began to be reported that this was much better than a holiday, as well as being a holiday in itself because it was so different. It was giving the women something they sorely needed: time out of danger, time to be looked after and free of responsibility, but also food for the mind and heart – understanding, inspiration, encouragement and love.

I cannot adequately express my respect for these women, for their courage, vitality and determination, their sheer will to keep going, keep living, keep their lives, their dignity and humanity intact. The way they went back to work after the bomb-scare was symbolic of the way they cope at 'home'; how they keep picking themselves up and doggedly carrying on with their lives. One of them described how the women of her home town, Sarajevo, managed, without chemist's shops, without even running water, to keep themselves immaculately turned out – not so much because it matters in itself, but because it symbolises their determination to retain their dignity in the midst of degradation. Others spoke with weary pride of the fact that every day they get up and prepare food, take their children to school, do whatever work there is to be done. This is the way they survive. And in addition they still find the will to keep struggling for the maintenance, or development, or resuscitation, of those things in society which even now embody human decency. This was the common ground they found, the spirit they shared and nourished in this week by the lake.

In the closing session our interpreter, a highly professional but somewhat unapproachable woman finally thawed and spoke for herself: 'I spend a lot of time at important conferences where the words are all empty – hot air – and nothing real is said. The contrast could not be greater with this workshop, where everything that has been said has come from the heart. I have been immeasurably enriched.' She spoke for me too. That communication of such depth and honesty took place in such a group and at such a time seems to me clear proof of the power of the human need and will to connect, of our instinctive recognition of our interdependence. Given our capacity for hurt and destruction, thank God for such a power and such an instinct.

CONCLUDING REFLECTIONS

As I indicated at the beginning of this chapter, I had had misgivings about agreeing to be the sole facilitator for this workshop and came to see that these qualms were justified. My role was too powerful and, by the same token, too responsible. At times I felt exhausted, and when Maria walked out of the session I had to stay with the group and had no co-facilitator to go after her. In terms of reviewing group dynamics and needs, I was greatly helped by Vera, but that was just lucky and not prearranged. I believe that both workshop facilitators and workshop funding agencies should take seriously the importance of co-facilitation.

At Balaton, I felt altogether welcomed by the group and had a clear, agreed role to fulfil. And although I felt afterwards that I had carried too much power and responsibility, I believe that at the time I managed it well and was of real service. Resistance to perceived Western interference and power was not an issue here and indeed I have never experienced any such dynamics when working with people from the former Yugoslavia. However, since the region now has so many excellent trainers and facilitators of its own, they rightly expect to be regarded as the primary resource in their own context. When I am asked to work there nowadays, it is most often because, as in Hungary, the task is to facilitate an encounter between people from competing or hostile ethnic groups. As I commented at the end of the last chapter, in such situations being an outsider helps one to be trusted to behave impartially.

Those who step outside the confines of prevailing inter-group mistrust and hostility are taking risks, both psychological and physical. It is true, as Maria said, that there could be spies in the group, though the workshop organisers can be expected to select participants with care; and it is always possible that what takes place in such a group could increase rather than diminish fear and hostility. At any rate, substantial emotional demands will be made on participants. My experience of many such workshops leads me to have a strong belief in the capacity of people who wish to do so to understand each other and form friendships across divides. I have never encountered hostility or anger that could not be managed and transcended. Participants come because they want to achieve something positive, and therefore they do. This does not, however, relieve facilitators of the responsibility to play their role as considerately and skilfully as possible, weighing the benefits and risks attached to the decisions they have to take. Part of the way they can contribute to the group's capacity to manage the conflict inherent in its make-up is themselves to behave with manifest respect for the needs and identities of all the participants.

When, in this account, I alluded to the vulnerability of the two Serb participants, not only because of their small number but also because of their lack of obvious victimhood, I was touching on one of the most interesting and difficult aspects of working with conflict in this kind of group. The power relations that prevail in the world outside are replaced by the 'parity of esteem' which is the hallmark of workshops for conflict transformation. Often numerical relationships between participants from different ethnic groups are also different from those experienced in their home context. In addition, the political or numerical vulnerability of a particular constituent group outside the workshop may be converted into a kind of moral dominance within the workshop. In this way the group and its facilitators have to work with a kind of inverted asymmetry. This dynamic will be further illustrated in Chapter 8.

It is often difficult to trace the after-effects of workshops, particularly if they are a 'one-off' event, as this one was. But just as later feedback provided evidence of the long-term usefulness of the Harare workshop, to several of its participants at least (and one hopes that this evidence is more widely indicative), later contacts revealed that the workshop held by Lake Balaton also had a lasting impact on some of its participants. I heard from both Maria and Olga, and their friendship continued. I met up with several other participants in later workshops and found that they were active in local peace networks and engaged in even more challenging dialogue work at home, for instance, in East Slavonia, when the large-scale fighting was over but inter-ethnic tensions were still acute. Vera had started an excellent dialogue project in that region, and assured me that it was our workshop that had made her believe she could do it. Another participant had set up a new peace group in her town.

The week spent with those people still has its impact on me. Like the Harare workshop, it deepened my determination that women, who in times of war hold societies together and have to deal with the dreadful effects of violence, should be recognised as having a vital role to play in public life, and that I should work to support them in taking their rightful place.

8 Surviving History: A Story of Dialogue in the Balkans

This chapter will focus on an ongoing process rather than a single event, one that continues at the time of writing and whose primary purpose, when it began, was not training but dialogue between different parties to a conflict. (As it evolved through changing circumstances, the programme's character became more mixed.) The example taken is of an unfolding series of workshops held before and since the war over Kosovo/a. I will describe, in varying detail, its different elements up to the time of writing, noting the impact of outside events on the process. The major shift necessitated by the war and its impact will be the occasion for reflection on the relationship between ideals, theory and political reality. It will be noted that, as suggested in Chapter 4, the methodology of dialogue workshops is often similar to that of training. Descriptions of the more recent workshops, especially, will demonstrate the overlapping functions of dialogue and training and their potential relationship to action.

Background to the Process

In 1987 Slobodan Milosevic came to power in Serbia, building his popularity on his nationalist rhetoric and using his defence of Serb interests in Kosovo/a as a political stepping stone. Denouncing claimed discrimination against Serbs by the Albanian majority in what was then an autonomous region, he rescinded Kosovo/a's autonomy. Albanians were dismissed from public service positions and a campaign of exclusion and harassment began. The Albanian response, under the leadership of Ibrahim Rugova and others, was to establish (with financial support from their diaspora) a system of parallel institutions for education, health and other services for their people. Other ethnic minorities remained, on the whole, within the official system and often were drafted in to fill the more menial positions from which Albanians had been dismissed.

In 1992, the Albanians organised their own elections for a president and a parliament, with some participation from other ethnic groups. Rugova was elected president and his party, the Democratic League of Kosova (LDK), overwhelmingly won these parliamentary elections. As Serbs and Montenegrins did not participate, some seats were held 'vacant' for them. The persecution, however, continued. Although, by and large, the 'international community' did very little in response, there was an attempt to broker negotiations on education during the 'Panic interlude' (late 1992–early 1993) and a small Conference for Security and Co-operation in Europe (CSCE) mission was sent. However, in June 1993 Slobodan Milosevic refused to renew its mandate because of the exclusion of the Federal Republic of Yugoslavia from the CSCE. Second, after Dayton, there remained an 'outer wall of sanctions' in place against FRY pending serious intent to negotiate about the status of Kosovo/a and respect human rights there. These, however, had no positive effect. Over the years, the majority population of Kosovo/a became increasingly weary, angry and frustrated, and some began to tire of the nonviolent nature of their campaign for independence. In 1996 the KLA was born. Its actions against Serb forces became the pretext for intensified repression and increasing international attention.

The dispute between Serbs and the KLA entered a new and dangerous phase in March–June 1998, with low-intensity fighting, destruction of villages and large-scale flight of refugees. In June the first workshop in this dialogue project took place. The idea for the project had been born the previous year in a small Quaker-founded institute within the politics and international relations department of an English university devoted to teaching and research on constructive approaches to conflict. The idea had its origins in the personal concerns of two former MA students at the institute, one Serb and the other Albanian (from Albania). In the event, only the latter was involved in the project's implementation, and then only in its early stages. Nonetheless, they were the catalysts for its inception. I was invited to work with the institute's director as a consultant to the project and (co-)facilitator of the workshops.

The dialogue project was not intended to address diplomatic issues at decision-making level, but rather to facilitate a process of dialogue at the level of students that, if successful, might progressively involve wider groups in each community. It was hoped that if young people and students from Belgrade and from the Albanian and Serb communities in Prishtina could find a better understanding among themselves of the issues in conflict, this could contribute to a wider movement for dialogue and against war. Such a dialogue might complement the movement for democratisation and civil society in Serbia, which were seen as prerequisites for political change. It was hoped that the experience of a conflict resolution workshop might challenge stereotypes and foster a willingness to accept the possibility of alternatives to war for managing the conflict. As the institute's director pointed out, the report of the International Crisis Group on the conflict had identified dialogue between students as one of the priorities for action and confidence-building measures:

> Positioned between two more extreme political alternatives ... the Kosovar students' movement may provide the best basis on which to build an effective, moderate opposition capable of putting forward a credible and peaceful plan of action .... They should also be encouraged to get in touch and collaborate with students from Belgrade.

The aim of the first workshop, held in Greece, was to bring together individuals from the parties in conflict, specifically student leaders and other socially or politically active young people, for a discussion on the long-term relationship between the two communities and the steps that might be taken in the shorter term to build confidence and trust between the two sides. Students' organisations have traditionally played an important role in both Serbia (where they took a leading part in the demonstrations against the Milosevic regime in 1996) and Kosovo/a (where they led the 1981 protests and the recent demonstrations for the re-opening of Prishtina University from October 1997).

We were aware of previous unofficial initiatives for dialogue, both at the higher and grass-roots levels. It was hoped that this workshop could complement and build on previous efforts and offer a renewed impetus to dialogue, with a view to strengthening and supporting the constituency for a peaceful outcome.

The idea of bringing together individuals from different parties in the conflict immediately raised the question of *which* parties. It became apparent that there were, even in the simplest framing (one which overlooked all minority ethnic groups in Serbia 'proper' and all but the Serb minority in Kosovo/a), three rather than two parties which had to be included. Those parties were not equidistant from one another and not comparable numerically. How were the asymmetries of these relationships to be reflected (or not) in the composition of the group invited to the workshop? Putting the demands of political realities and sensitivities in the scales with those of group dynamics and power relations within the proposed workshop, we invited six Albanians from Kosovo/a, three Serbs from Kosovo/a and three Serbs from 'Serbia proper'.

My spelling formula for 'Kosovo/a' and the awkward and potentially insulting 'Serbia proper' indicate another dilemma entailed in the workshop's framing, which quickly confronted us as would-be facilitators of dialogue. How could we talk with the participants about territories whose relationship to each other was the point at issue between them, without positioning ourselves with one side or the other by our very choice of language when referring to those territories? In practice we fumbled our way forward with explanations of our dilemma, asking participants and partners to forgive any offence we might give. (The institute, though not always its facilitators, in fact stayed with the official, internationally accepted spelling, 'Kosovo', but the official international position that this reflects has become increasingly distant from the tacitly acknowledged reality.)

The local partner for the project in Belgrade was a student organisation, with branches in several cities, committed to human rights, intercultural and international co-operation and understanding, nonviolent communication and conflict resolution. On the Albanian Kosovar side, we approached a mental health association based in Prishtina whose co-ordinators were previously known to me and had expressed an interest in helping to enable political dialogue. It was, at the time, one of the few NGOs in Kosovo/a which maintained active contacts with both Albanians and Serbs, through its mental health work and 'round tables'. These two organisations were involved in all the planning, as well as identifying enthusiastic participants and acting as local organisers.

First Joint Workshop: A Small Breakthrough

The first workshop was held at Loutraki(s) in Thessaloniki/a (other politically sensitive names). The group assembled was composed in the way intended, except that in the event there were only two Serbs from Kosovo/a. All the participants had travelled together from Prishtina, which was in itself promising, but they were very tense when we met for our first session. Having taken care to make adequate space for them to express their hopes and anxieties in relation to the workshop (and it helped that these were similar on all sides), and made clear agreements about the process of the workshop (including confidentiality), we plunged straight into the heart of the conflict by asking the three different groupings to work separately and formulate what they thought was at issue in the conflict and what should be done: in other words, to present their positions. In the first place they resisted the idea of being separated out in this way so that their differences were exposed, but they allowed themselves to be persuaded that this was necessary if we were ever to make progress in relation to what lay beneath the surface and was dividing them whether they liked it or not.

Having accepted that this was a task to be tackled, the different groups worked hard and made very clear presentations, which were therefore quite inflammatory, and a heated discussion followed. It was interesting to note that the Albanian Kosovars presented a united and verbally pugnacious front, brooking no implied criticism of the KLA, whereas the Belgrade Serbs were softer and more ready to concede wrongs on 'their' side. The Prishtina Serbs were conceding and almost pleading, only defending their right to live in Kosovo/a, and that quite movingly. The dynamics of this exchange represented a reversal of external power relations, reflecting the implicit acceptance by all that the Albanian Kosovars were occupying the moral high ground.

The hostility reflected and generated by this exchange was transformed almost miraculously by the subsequent exercise, undertaken in the same groups, in which each expressed the needs and fears of its members in relation to the conflict and its potential outcome. Not only were some of the needs and fears the same, but all were so recognisable and undeniably understandable that their presentation created empathy in every direction. It was particularly striking that the Serbs from Kosovo/a now had a voice that was heard. And so power relations in the group not only shifted but became less important.

The next exercise undertaken was a role-play of a particular micro conflict, the disputed use of the Medical Faculty buildings at Prishtina University. A reverse-role role-play of mediated negotiations produced not only new insights but some rather practical joint proposals, which were further developed by one working group comprising members of all three sub-groups, while another working group discussed ways forward on the macro conflict over the constitutional status of Kosovo/a. The participants discussing the use of the Medical Faculty buildings were keen to involve the institute in a mediation process, to work towards a resolution of the conflict. They said they would contact student leaders and faculty staff when they got home. Those whose focus was the wider political conflict had a more intractable problem to address. They proposed a process for developing dialogue on the constitutional status of Kosovo/a and were willing to be active in it.

During the five days of the workshop, the different group members bonded socially and drew much closer to each other in terms of their understanding of the situation confronting them all. Firm friendships were made across ethnic lines, and specific plans were drawn up for an attempt at mediating student negotiations about the use of the Medical Faculty buildings, with two of the participants undertaking to contact friends and colleagues with a view to forwarding the process, and the institute agreeing to provide facilitators and seek funding for it. It was seen both as a small enough project to be within our compass but at the same time as having significance in relation to the larger political conflict of which it was emblematic.

Second Joint Workshop: Under the Shadow

The second Serb–Albanian Kosovo/a dialogue workshop was held in Bristol in February 1999. (It is perhaps instructive to note the number of months it had taken to organise it and the logistical difficulties which caused the gap to be so long. The time-consuming nature of preparation for such events had been discovered to necessitate the appointment of a part-time worker. Funding proposals had taken a long time to prepare and responses were slow in coming.) No real movement on the Medical Faculty issue had been achieved, in the absence of any meetings since the first workshop. While it was hoped that this second workshop would be able to pick up the issue and maybe take it further, by this time the situation in Kosovo/a was even worse. OSCE monitors had been sent to the region and had had some impact, but murders were continuing and the Rambouillet 'talks' were about to begin. We gathered with a sense of nervous expectancy. We were determined to take our own process seriously, at the same time being aware that a larger process, beyond our control, was underway and would be decisive for the region, and that almost anything we decided would be dependent in some way on the outcome of that process.

The decision had been taken to mix former participants with new ones, to provide continuity and the possibility of building on the Thessaloniki/a workshop, and at the same time to draw in more people. While still recognising the good sense of this policy, I also believe it had disadvantages, since former participants were ready to move forward while new ones needed to cover much of the original ground. This time I worked with a new co-facilitator who has continued as a colleague in this work. Together we devised a process which included (in addition to the usual introductory group-building elements) sharing visions for the future of the region and dialogue about the application of these visions to the question of Kosovo/a's political status. The discussion was heated but never hostile. (Here I think we did benefit greatly from the experience of the first workshop and the trust it had generated.) At one stage we used a 'fishbowl' discussion process to slow the process down (with discussion limited to the interaction of four participants at a time, who sat in the centre and could be replaced at any time by others who wished to speak). In the end, however, this braking mechanism became itself the object of collective hostility and was abandoned in favour of a healthy (facilitated) free for all. No agreement on the constitutional issue was reached, except that the group was willing to go on talking about it, admitting that it was possible to do so and by implication that nothing was set in stone, on either side.

The proposed mediation process in relation to the Medical Faculty buildings was again discussed, this time in an even more detailed and practical way, and corresponding plans were laid, but it was acknowledged that these plans would make sense only if there was some kind of positive outcome to the Rambouillet talks.

Other plans were made for small-scale activities in Prishtina, trying to create some common space for social interaction and dialogue (in the first place on 'safe' topics) and in support of the responsible and inclusive use of the media. (We were lucky enough to have the participation of two committed young journalists from the 'free' newspaper, *Koha Ditore*.) But here again we were all aware that such plans had meaning only in the context of a positive outcome to what was going on in Rambouillet. It seemed both vital and a little ridiculous to be making such plans at such a time, epitomising the paradoxical need to be working for political participation in a world where political power is currently held by the few.

The War

In the event, the outcome of the Rambouillet process was (it now seems predictably) not progress towards some kind of peace with some kind of justice, but a massive bombing onslaught by NATO against Serbia, and an unchecked and massive explosion of atrocities by the Serb army and militias against the Albanian population of

Kosovo/a, in the face of which the continuing activity of the KLA (which had played such a role in accelerating the crisis) was more or less impotent. A vast proportion of the population of Kosovo/a took refuge in Albania, Macedonia and elsewhere. Many Serb activists also left Belgrade for Budapest and other destinations. In that time e-mail provided a vital link between former participants with the institute. By the time the bombing was over and KFOR (Kosovo Force) was entering Kosovo/a, it felt as if these tenuous connections between scattered and traumatised people were all that was left of the dialogue process.

In addition, the shape, composition and nature of the conflict had been radically altered, which meant that the dialogue process that had been begun might not fit the new situation. In the first place, a new party, NATO (succeeded, ostensibly, by the UN and KFOR), had entered the conflict and assumed the dominant role. Secondly, as a consequence, power relations had been reversed, and the Albanian population of Kosovo/a were cast in the role of victors, while the rest of Serbia had lost any influence in what had been its province, had been severely damaged and further impoverished, and (despite Milosevic's victory claims) was in political disarray.

The advent of a third party into the conflict had additional significance for the dialogue process, in that the institute, as part of a British university, was now liable to be identified with that party and its actions, and therefore no longer regarded as impartial. At the same time, those of us employed by the institute had all opposed the bombing, knowing that some if not all of our Kosovar Albanian participants were likely to have supported it, more or less reluctantly.

From the point of view of those of us who had been involved in the institute's facilitative work, this whole period, while of course nothing like the nightmare it was for our participants from the region, was extremely distressing. It seemed important to express the human solidarity we felt for all our participants. It also seemed vital to reconnect with partners and participants whose whereabouts we knew. It was therefore decided that my colleague and I should go to Macedonia, as representatives of the institute, and meet with those participants from Kosovo/a, both Serb and Albanian, who were at that time in Macedonia, in Tetovo ('Tetova' to its majority of ethnic Albanian inhabitants) and Skopje. Our purposes were to express our concern and commitment and to learn the current feelings and hopes of those we met, how their aspirations could be supported and in particular their views about the possibilities and need for dialogue in the changed situation. We were accompanied by a member of an international peace team which was then working in Kosovo/a who knew several of our former participants, had excellent contacts in the region and had been extremely helpful to us in arranging the visit.

Visit to Macedonia

Because in the event our visit to Macedonia (in mid-June, 1999) coincided with the arrival of the first KFOR troops in Kosovo/a, and several former participants in the workshops were working for the OSCE and were frenetically busy during those days, we could not meet with them all at once or for a substantial amount of time. We met for a day with a very small group of Albanian Kosovars whom we had never met before, but who proved very interested and interesting; for an evening with our partners from Prishtina; and for part of a morning with a small mixed group of former participants, Serb and Albanian (together with one new person). These conversations gave us some idea of what they had experienced, and of the feelings, on the Albanian side, of both horror and triumph at the turn events had taken. It was clear that it was the opinion of most of them that now was not a time for dialogue with a defeated people but for recovery in their own community.

One Albanian Kosovar we met, who had participated in both the Loutraki(s) and Bristol workshops, reaffirmed her willingness to stay with the dialogue process and to meet with Serbs as soon as could be arranged, but even she questioned why this bilateral relationship with Serbia should still be recognised. This was a strong sign for us, as we could have guessed from experience elsewhere (for instance, relations between the Abkhaz and Georgia), that the *de facto* separation of Kosovo/a from Serbia would result in a wish on the Albanian side of the dialogue to deny any special relationship to the country which had so very recently denied its majority population civil rights, let alone autonomy or independence.

On the other hand, and equally understandably, the Serbs we met were eager for the dialogue to continue and could not see that anything that had happened should change the attitude of their former dialogue partners to its continuation. They identified themselves as the new underdogs politically, but thought that within our dialogue process this reversal should make no difference.

Although these meetings in Macedonia were so small and *ad hoc*, they were vital to our understanding of the current realities and feelings of friends and colleagues, and what was and was not currently needed or possible in the continuation of the dialogue project. We also felt for ourselves the inter-ethnic tension in Macedonia where, according to what we were told, important and positive efforts were being made at the governmental level to improve things, but where relations on the ground between ethnic Slavs and ethnic Albanians were often extremely hostile and volatile.

As a consequence of this visit, it was decided that at this stage in the unfolding regional conflict, after all the traumatic and cruel events of the past months, it was not possible to bring Serbs from 'Serbia proper' together with Albanians from Kosovo/a, and that the most appropriate role for mediatory facilitators in a situation of such shattered relationships was to work with different parties separately, to help them process what had happened and look again at what was possible for them to do in their own situations, and discover what support was needed. Plans were therefore laid for us to travel to Belgrade. Those invited to participate were not only participants from the earlier workshops, but also other NGO activists I knew. I was diffident about inviting them, aware that we would be taking up their time without, perhaps, offering them anything in return, but they were reassuring and seemed to value the idea of such a gathering and the chance it would offer for concerted thinking.

Workshop in Belgrade

Having met in Macedonia with some of our participants from Kosovo/a, we decided it was important also to visit those who lived in Belgrade and arrangements were duly made. Our contact in the partner organisation there had left for Budapest, but we were assisted in making our arrangements by a colleague of his who had remained, and by another organisation with which we had connections, who invited a dozen individuals to a two-day workshop with us.

It felt strange and uncomfortable to be going as facilitators to a country which had so recently been bombed by our own. Although we had been prepared for some hostility on our journey from Budapest to Belgrade, we encountered none. We were, however, stunned by the sight of broken bridges and burned-out buildings, however much we had expected them.

We began with an evening meeting with former participants. As it had been with those we had met in Macedonia, it was moving to hear what they had experienced since we had last met. One of them was now a refugee from Prishtina, who felt he would never be able to return and spoke of the terrible days before he had left.

We spent the first half-day of the workshop exploring how the participants (all of them known to us but not all as part of the dialogue programme) felt about their current situation. While their aspirations seemed unchanged, a great deal of despair was expressed: a sense of hopelessness and exhaustion and lack of motivation. Hurt was voiced by several that former friends had seemingly turned against them and had supported the bombing and ostracising of Serbia. There was general agreement that without the removal of the Milosevic government no radical improvement in the situation could be expected. The two main lines of thinking, pursued in successive rounds of group work and plenary discussion, were the reaffirmation of the purpose and meaning of holding on to cherished values and working for change, and ways in which the movement for human rights, peace and democracy in Serbia could become more co-operative, inclusive and effective. The combination of the two seemed to restore some of the energy and sense of focus which participants had been missing.

Towards the end of the workshop, when we were beginning to explore the question of dialogue – with whom it was needed and for what purpose – we used a group 'sculpting' process to feel and discuss how regional relationships were perceived. What emerged was a sense of isolation and victimhood within the region, with the 'international community' playing a largely negative role towards Serbia and influencing its relations with its neighbours. Little mention was made of inter-ethnic relations within Serbia itself. There was great eagerness for the renewal of contacts with Albanian Kosovars, and a recognition that intimidation from their own people might make former friends or colleagues unwilling to associate with them. The idea of a regional workshop, to include 'Serbia proper', Kosovo/a, Macedonia and Montenegro, was seen as a possible alternative if bilateral dialogue proved not to be feasible. In the light of the nature of concerns and issues raised in this workshop, we felt that the meaning of 'civil society' and 'multi-ethnic democracy' would be the most pertinent topic for regional dialogue.

During this visit it became clear that our former partners in Belgrade were depleted in membership and unclear as to their future. We would therefore need to build on other local partnerships.

We (the institute team) had been increasingly aware of the exclusiveness of a Serb–Albanian axis, and in particular of the absence, hitherto, of Roma groupings from our thinking and our workshops. We therefore used the good offices of a Belgrade colleague to meet with two Roma leaders, to explain our project and our awareness of its exclusiveness and explore their interest in future participation. One of them seemed interested only in advocacy and development for Roma, but the other understood that we were talking about something different and was keen to participate, offering to find suitable participants for a future regional workshop.

The obvious next step was a comparable visit to Prishtina, now that most refugees had returned, including our partner organisation there. (Also, none of the institute's team had ever been to Kosovo/a, which seemed a glaring gap in our experience.) The purposes of this visit would be the same as those for Belgrade, and framed in a similar way. We had gone ahead with the Belgrade visit without seeking permission from our governmental funding agency to use funds intended for bipartisan dialogue for work with one group only. We therefore needed to justify our action and to submit a plan for visiting Prishtina. This again took time, and it was only a matter of days before the planned visit when we got the final go-ahead. This in turn meant that our local partners (helped by the peace team mentioned above) had little time for confirming invitations, and had held back in approaching potential workshop participants. (I mention these logistical difficulties because they are an inevitable and important aspect of such work and need to be taken into account and learned from, so that everything possible can be done to obviate their impact.) In addition, my regular colleague was ill and forbidden to travel, and he and I could find no reasonable alternative dates for travelling together. Fortunately, a staff member from Quaker Peace and Service was able to step into the breach.

Workshop in Prishtina

As we had been stunned by the sight of the damage inflicted on Serbia by NATO bombs, so now, on the car journey from Skopje to Prishtina, we were sickened to see the mass graves and burned-out homes – the result of Serb military action against Albanian Kosovars. We were also struck by the omnipresence of 'internationals', conspicuously in KFOR, Red Cross, OSCE and other such vehicles. As our driver said ironically, 'Welcome to free Kosova!'

The workshop of the next two days proved a small one, with most of the eight or so ethnic Albanian participants absent at certain times because they had urgent tasks to attend to as part of their paid employment. In spite of this disruption, we were all, I think – facilitators, participants and local partners – gratified by the quality of thinking and the intensity and openness of the exchange. The tone of openness was set in the introductory round and was surprising, especially since what emerged in all that followed was the pervasive pressure within the Albanian Kosovar population against speaking out. All our participants seemed to agree that to contradict nationalist triumphalism, and the ethnic and linguistic bigotry associated with it, was to risk social and indeed physical attack. They lived, some claimed and others admitted, in an atmosphere of intimidation. It was also said that the two or three months of the bombing were worse than the previous ten years put together. At the same time, there was an undoubted sense of liberation and possibility – exhilaration almost – and a sense of unreality. 'We need to get our feet back on the ground', someone said. The participants were aware that at present they were almost totally reliant on the 'internationals', who had brought temporary prosperity to the young and educated in Prishtina, while others in the city and elsewhere had no adequate means of earning their living, and local structures and services were in ruins.

Most of the workshop was spent addressing the need to overcome inter-ethnic hatred and discrimination and to restore the space for human interaction without ethnic labels. Acting to counter the effects of intimidation against free speech was seen as a vital precondition for this. The role of the media, especially newspapers and radio, was emphasised.

In the final afternoon, as we had in Belgrade, we explored the group's view of relationships in the region (again using 'sculpting' to bring the discussion to life), and appropriate contexts for dialogue. The isolation of Serbia, which had characterised the Belgrade 'sculptures', was mirrored here. Yet participants recognised that, in the long run, the different units within their region were interdependent, and that to isolate one member of that geographical collective perpetually was untenable and dangerous. They were not prepared to be party to bipartisan dialogue with Serbs from Serbia, since to acknowledge a particular relationship with them would be to undermine progress towards an internationally recognised state of Kosova. They acknowledged the claims of Serb refugees from Kosovo/a and the great need for inter-ethnic dialogue within Kosovo/a itself. It was said that in particular situations, in towns or villages where significant numbers of non-Albanians were still present and tension and fighting were making daily life stressful and dangerous, dialogue mediated by outsiders was a necessity, albeit a risky one. However, it was felt to be too soon for more general dialogue with Serbs and others. (The others mentioned were Montenegrins, Bosniacs – seen to be currently even more at risk than Serbs – Gorani and Turks. When we asked about Roma, there was hesitant and apparently grudging agreement that they too needed to be included on the list.)

Regional dialogue, on a common issue other than constitutional relationships, was seen to be both realistic and desirable to aim for. After some discussion, it was agreed that a workshop for such dialogue should include participants from Kosovo/a, 'Serbia proper', Montenegro and Macedonia. Albania was suggested, but at that time agreed to be potentially problematic in the way its inclusion would be perceived by others (that is, threatening to Serbs); likewise Bosnia and Croatia. Our restricted list seemed one that could make the dialogue possible for all its constituents. When we came to a discussion of possible venues, there were two considerations: access (visas) and neutrality. Obtaining travel documents was identified as a potential problem for Kosovars, most of whom had had theirs confiscated. The topic proposed unanimously for this regional workshop was civil society and the establishment of democratic, multi-ethnic democracy.

At the end of the workshop it was acknowledged that it had provided an important space for its participants to speak freely and think concertedly about the new situation into which the participants had been so traumatically and suddenly plunged. It had given us facilitators a chance to learn about the perceptions of a few individuals who could reasonably be thought to be in some way representative of Albanians in Kosovo/a who were trying to hold on to the values of human equality and co-existence. We had also, in the process, been given a great deal of information about events and relationships in other places and social strata. The disappointment was the absence from this workshop of former participants in the dialogue process. Some were prevented by work and others could not be contacted (though we met up with one of them accidentally). Not one of the Prishtina Serbs who had participated in the past was now living in Kosovo/a. This added to the feeling that the whole process had been blown apart by events and that picking up the threads of continuity was going to be difficult, if indeed to any appreciable degree possible.

Reframing the Dialogue

In the light of our visits to Macedonia, Belgrade and Prishtina, we were left in no doubt that the dialogue that had begun as a complex bilateral one had now been forced by events into a multilateral, regional frame. It was questionable to what degree this could be seen as a continuation, rather than as a new process. Our former Serb participants from Prishtina were gone, though one was living in Belgrade and therefore potentially able to be included still. The Belgrade and Prishtina workshops had involved new, very good participants who had shown interest in involvement and whom we would want to include in future activities. Widening the geographical frame would mean that very few from each constituency would be able to be invited to a given workshop. (Quite apart from questions of ethnic 'balance' and political sensitivities, this was one reason for limiting the regional frame to the four units mentioned.) This would exacerbate the old tension between the need for continuity and the need for expansion. It would also make it difficult to include enough participants from any one part of the region to provide them with mutual support within the workshop and on their return home.

There was also the question of just what mix of participants was to be drawn from the different parts of the region in question, each of which contained a number of ethnic groups. Given the level of tension and intimidation then current in Kosovo/a, it was judged to be wise, in the first instance, to limit participation to four members of the majority Albanian population, though possibly inviting Serb refugees now living outside the territory, as well as Serbs from 'Serbia proper', to equal their number. Since the plight of Roma people in the whole region was both so urgent and so particularly difficult to address in this context, we felt it was important to try and include at least two Roma participants from Serbia, where we had the contacts established during our visit to Belgrade. To try to ensure that they were neither victimised by others nor adopted a simply victim role in the workshop, we would stress the desirability of identifying two articulate and confident young people with an interest in dialogue, as well as self-advocacy. From Macedonia we would invite two Slavs and two Albanians. At that time we had few contacts in Montenegro but planned to try to find at least one or two participants from there.

This list seemed to give an acceptable mix of potential participation, not too widely (and therefore thinly) spread. I say 'seemed'. It was a question of feel, rather than arithmetic or science, though arithmetic was necessary to ensure that we did not exceed numbers which were financially possible or manageable in terms of group dynamics. I have laid out these minutiae to illustrate the complexities and tensions involved in planning such a workshop: one which would in this case, we recognised, exclude many groups which are present in the region. There are so many factors to be borne in mind – resources, external realities and group dynamics, sensitivities and symbolism, self-perceptions within groups, and the way others perceive them. Through it all runs the thread of power in relationships – including the power of the given Western organisation which has the position and the access to funds to orchestrate such events. The fine-tuning of participants' lists, for an event costing thousands of pounds, holding the power to include or exclude, to write down or cross out, can (and perhaps should) feel indecent. It reflects the power relations which characterise the world, from which we cannot escape, except by working within them to change them. I was reminded in this process of the need to stay awake to the associated discomfort, sharing power in every way possible and trying always to listen to the opinions and serve the needs of those with whom we were working.

First Regional Workshop on Democracy

The first regional workshop (the last workshop for which funding was in place) was held in Bulgaria in December 1999. It included some participants from previous workshops and, inevitably, given the new geographical scope, a large proportion of new ones. In the event, there were three Albanian participants from Kosovo/a, three Serbs and two Roma from Serbia, one Montenegrin, and two Slavs and two Albanians from Macedonia. There was a fair balance (as there had been in all the preceding workshops) between male and female participants. They comprised student leaders, journalists, NGO workers and educators. The presence of a few 'old timers' in the group provided a degree of confidence that this experiment could work, which was important in view of the confessed nervousness of many of the participants. When they talked about their hopes, these centred around the possibility of new and common ways of seeing things, the development of new projects that could really be implemented and the creation of new contacts and friendships with members of other ethnic groups. The fears expressed were of being unable to understand each other's sufferings; of concealing the 'real problems' (that is, political), or, when 'real problems' came up, of being unable to deal with them together; and, finally, of not being understood by family, friends and colleagues when they went home with new ideas.

In terms of the immediate dynamics within the group, the participants quickly relaxed. However, after a morning's work on the question of identity and representation, they realised that there was a contradiction between their desire to focus on the future and a need to talk together about what they had all, in very different ways, just gone through. It quickly became clear that there could be no real moving on together if this need was ignored, so we proposed a 'round', in which each participant had the chance to speak into a pool of silence, with a pause between each contribution. This proved to be a very moving exercise. The stories were very different but the pain of them was palpable and shared, and the power of this sharing greatly altered mutual perceptions. We returned to the questions of identity, representation and democracy with a much deeper basis of understanding from which to work.

On the third day, the focus of the discussion shifted to civil society and organisational democracy. This focus led, in turn, to an exploration of different models of leadership, which was developed through two substantial and well-processed role-plays. The enthusiasm which these engendered was followed by a demand that they should use role-play to address 'real' topics, so the next day found the participants analysing the Kosovo/a–Serbia conflict, first in its macro form and secondly as embodied in the relationship between a Serb wife born and bred in Kosovo/a and her Albanian Kosovar husband. This latter, personal, version of the conflict was played out in a mediation role-play which involved and touched all the participants, whether as players or as observers.

On the last full day we turned our attention to potential actions. Two ideas were explored, one being a 'twinning' scheme for schools and the other an approach to local media to try and generate support for the organisation of a music festival for peace in the Southern Balkans region. These ideas were developed with enthusiasm, but it was clear that, at that time, the participants lacked the confidence and resources to turn these ideas into serious plans and funding proposals, so that they could implement them.

The final day started with a tired and emotional group of participants, many of whom had been enjoying the local night life or conversing in their rooms until the early hours of the morning. As the workshop drew to a close, the participants expressed their great satisfaction at the changes it had brought about in them. These changes were mainly related to the 'perception of the other' and the 'lessons for peace' that had been learnt by understanding more about conflict. Participants from the Roma community were very pleased to have met Albanians from Kosovo/a for the first time in their lives. Serb participants claimed that they were 'happy to see urban educated Albanian women because, before, all we could imagine were women from villages wearing head scarves'. (My colleague and I squirmed a little at this, noting its élitist (and sexist) tenor, but recognising at the same time that we were in fact concentrating our own efforts on the educated and relatively privileged members of the different ethnic groups.) Albanian participants from Kosovo/a said they had come to understand that Serbs were also victims of the war and not just 'guys with guns'. These new understandings had been possible, they said, because we had created an environment in which the participants had felt safe to express themselves.

From these reflections it was clear that it was the encounter with 'the other' that had been of paramount importance to the participants. It was also clear that even in this complex regional process the bipartisan dialogue with which the process had begun remained dominant.

We knew from their daily evaluations that the training element of the workshop had been appreciated, but it was the sense of inter-ethnic community that was, in the last analysis, most important to participants. They saw it as their responsibility to tell their families, relatives, friends and colleagues about the problem-solving approach to conflict to which they had been introduced in the workshop, and about the longing for peace which had been shared by the repre-sentatives of different ethnic communities. They said they were 'charged full of positive energy after feeling connected to a new circle of friends who shared the same fears and needs'. At the same time they recognised the courage and determination that would be needed to change the perceptions of others. Feelings of hostility were strong at home, and it was very likely that some of their new ideas would not be understood. They did, however, feel they had some new ideas for ways of approaching people with hard-line positions: to show their photographs to friends and colleagues, explaining the content of the workshop; to write articles for local newspapers and broadcast programmes using multi-ethnic radio stations like *Radio Kontact* in Prishtina; to sing Albanian songs with Serb children from

Belgrade; to talk about the workshop at other seminars; and to invite Albanian speakers to educational events in Belgrade. Networking between those working in NGOs was seen as vital for the development of common plans (for instance the creation of a multi-ethnic radio station in Mitrovica).

It was considered important to hold further workshops aimed at enlarging the representation and increasing the number of people involved in the process and resulting network. An Albanian participant from Kosovo/a suggested the participation of former KLA members who had experienced the war but now gone back to ordinary life, and also the inclusion of Serbs and Roma from Kosovo/a who had been affected by the war. (This reminded us of the restricted nature of our existing group.) It was also felt that to include participants from Albania would be both acceptable and, given that country's isolation, desirable.

The group parted with sadness and a determination to keep their community alive, as well as to act in whatever way they could to overcome intolerance and enmity at home. Much correspondence ensued, facilitated by the setting up of an e-mail list and a discussion/action web page.

Second Regional Workshop

In line with our participants' suggestions, a second regional workshop was organised for April 2000, at the same Bulgarian venue, drawing in a few additional participants, including a woman from Montenegro, a Serb Kosovar journalist from Mitrovica and two participants from Albania. By now, effectively, our participants had formed a network of partners and, though we stayed in touch with our original partners, it was they and their organisations who provided most of our new contacts.

This second workshop followed a similar pattern to the first, in terms of content, though the pattern of dynamics was different. The immediate rawness of postwar feelings had passed, and had been replaced by a different kind of wariness and suspicion. On this occasion the boil was lanced when my co-facilitator and I were challenged to reveal more of ourselves and our own perspectives, and took the risk of speaking openly about our own beliefs and values – including our opposition to war as a means of solving conflict, and in particular to the bombing of Serbia by NATO. It had been a matter of considerable discomfort, since those events, to know that some of our participants had been very glad of the bombing and would assume that we had supported it. Remaining silent felt uncomfortably like dissembling, especially when we had made it known in Belgrade that we had opposed the NATO action. At the same time we had not felt that our views were what was at issue within the dialogue process and had wanted to keep them out of the picture. Now, however, we found that our sincerity and openness were accepted and, far from generating mistrust or resentment, opened the way for a hard but not destructive political discussion, mainly between Albanian Kosovars and Belgrade Serbs, which clearly felt like a substantial achievement for the group.

Once again, analytical work led to plans for joint action, but did not seem likely to lead to implementation, since, as before, that would require further, more detailed planning and the writing of funding proposals – which in turn would involve more meetings for the geographically scattered planners, for which there was no money available. But participants were clear that they wanted the network to develop and to undertake some form of action.

Plans for Further Development of the Network

To go some way towards addressing the problem (for a project spread thinly across several countries) of achieving anything approaching 'critical mass', or an adequate degree of mutual support for the participants at home, so that action might become possible, three more workshops were planned for the year 2000. The first two of these were to involve a new set of participants (recommended by existing contacts) of similar size and composition to the first, introducing them to the same ideas and processes. The third – a larger one – was to bring the two new participant groups together, along with old participants, so that they would form one enlarged network. At the end of this last workshop, there would be a review and evaluation of the entire process, on the basis not only of the events themselves, and the understandings they had produced, but also of the activities and connections which had resulted.

The intended outcome was to have in place a well-established, regional network of young people (facilitated by a shared website) and a commitment not only to further developing those links but to pursuing and instigating local activities for interethnic tolerance, co-operation and democracy. In this way the project would contribute to the development of personal and organisational capacities for the establishment of a healthy civil society and democratic culture in the region.

Challenges of Maintaining Continuity

At this point, however, it was time to look for further funding if the programme was to be continued. The director of the institute which had so far been responsible, pressed by ever-growing commitments, now wanted to hand the programme on to another organisation and, through the good offices of a co-operative committee, the transfer to a London-based NGO was made. My co-facilitator and I were part of the transfer, and I became programme manager. When we applied for a grant for a further two workshops, however, funds were no longer available from the government ministry in question, whose regional focus had shifted, and many other potential sources refused our requests.

Eventually, an approach was made to the democracy-promoting foundation which had made the very first workshop possible. As luck would have it, they themselves had received belated funding for a regional programme of workshops for young people in the Balkans – a programme which they were no longer in a position to implement directly. Instead of our request for funding for three workshops being agreed or rejected, we were asked if we would become the implementing organisation for a much more extensive programme. We said we would need to think how our existing participants could benefit from this proposal and how they might help us to implement it, and that we would have to reshape it to fit with their needs and energies. Then we left the meeting in a daze.

We were in something of a quandary. On the one hand it was an exhilarating prospect to have the freedom to (re)design and implement, with our local partners, a substantial programme, and to have the resources to enable them to move from thinking about action to action itself. On the other hand, the implementation time would be less than a year. To work under such pressure, committing our existing and potential participants to organising events and projects without any certainty that they would want to take this opportunity, seemed almost irresponsibly risky. Yet to refuse the opportunity which this represented to enable them to meet again, extend their network and carry out some of their ideas would seem equally irresponsible. We had very little time in which to consult such a complex constituency and had to make a decision. We came down in favour of accepting the challenge.

We proposed to organise two initial seminars, with old and new participants from the same geographical region, in which participants would have the opportunity for some broad dialogue and then the chance to develop projects 'for real', with outline funding proposals backed by personal commitments to their implementation. The proposals would be finalised and agreed afterwards by e-mail and our organisation would provide the link with the funding foundation, plus advice and support for the projects' implementation, including visits to all of them while they were in progress. In addition, three training workshops would be held for those implementing the projects. At a final seminar, they would all meet to report to each other about their work, reflect on and learn from it, and think about its continuation, as well as about the future of the network.

This broad proposal, a modification of our earlier one, with a whole new layer of local activities implemented by participants, was accepted. As I write, seven resulting projects are under way, some local and some regional: a three-city, multi-ethnic series of workshops for young people in Serbia; a regional art competition, to be followed by a postcard campaign, with poems in all regional languages accompanied by English translation; a regional workshop on gender, sexuality and inter-ethnic relations and a regional, multi-ethnic summer camp; a children's multi-ethnic play scheme in Prishtina and a project for educating educators to combat the culture of violence among young people in Albania; and a regional project to make and distribute a film on human rights.

These are exciting results, but the workshops from which these projects emerged had a daunting number of goals and were difficult to facilitate. While there was external pressure to 'produce results' in the form of viable projects, we as facilitators were determined that participants should not feel obliged to make commitments for which they had no time or energy, or for which they lacked sufficient motivation. To meet, establish or re-establish trust and a real sense of connection, exchange views and information, share values and visions, transcend differences, establish common ground *and* make realistic plans together in the course of five days was, to say the least, a tall order. In some cases original plans needed to be let go of and new, more grounded ones developed. Those who were familiar with training workshops, or indeed dialogue workshops, found the necessarily open process we used unfamiliar and felt at first that this was not a proper workshop at all. Nonetheless, the final evaluation was, in both cases, enthusiastic. Participants felt they had exceeded their own expectations in what they had been able to achieve and that they had built strong bridges between them through their joint work, not only learning *about* project development but, for the first time in many cases, *doing* it. The resulting projects will teach them even more in the doing, and will provide learning opportunities for many more people in much wider circles.

CONCLUDING REFLECTIONS

I will, of necessity, end the story at this point. It has been long and complicated, but it continues. It has changed its shape almost beyond recognition, having begun in circumstances very different from those which now apply. A few original participants remain in the network, but most are new; yet the story is important to all of them. My first reflection, then, is that those who embark on work of this kind cannot know where it will lead and will need to be responsive as well as 'proactive'. They will need to know that they are responding to needs and desires 'on the ground' and establish a strong two-way relationship with partners – though partnerships, too, may be changed by changing circumstances. Providing continuity is vital, especially in the midst of discontinuity and dislocation, but flexibility in the way this is done may be the key to a process's survival.

My second reflection is that in all of these workshops, whether pre- or postwar, the skills and strategies of *both* nonviolence *and* conflict resolution were relevant. In the world as it is, those who work for peace, justice and democracy will always be struggling against strong opposing forces. And they will always have the task of building bridges and upholding respect and humanity in the face of enmity and disregard.

My third reflection is that even when macro events seem to have overriding power for discontinuity and dislocation, individuals and movements do have the power to continue in their chosen direction and will take into their new lives the things they learned and the intentions they formed in other times. Several participants in those prewar seminars, while they are no longer active in this network, are still travelling along parallel roads in other organisations and countries. They will have their effect in whatever spheres they live their lives. And meanwhile the work they began is continued by others.

Events in the Balkans have been extremely discouraging in recent years. Untold numbers have died, and life for a high proportion of survivors has been made wretched. Inter-ethnic relations in some places are not improving but deteriorating, and peace and democracy can seem like a distant dream. The world of ordinary people – what Ropers (1995) calls the realm of societies – can be shaken and greatly changed by political events planned in the realm of states. Yet these realms are constructs whose boundaries are transcended by individual human beings who act, even in the most adverse of circumstances, with some degree of autonomy, however limited and unpredictable. The will and capacities of those individuals constitute threads of continuity which are woven between and across the threads of events and are as real as anything else in any situation.

My fourth reflection is that there are all kinds of complexities involved in preparing the kinds of events described that cannot be addressed in single or simple ways, by some formula, but have to be somehow weighed and fashioned into reasonable shape. Organising and facilitating them is therefore more of a craft than a science. The design of a programme needs to hold together the logic of external realities with the needs of group dynamics, which will spring from those external realities but not necessarily mirror them. There may be competing needs for continuity on the one hand and expansion of the circle on the other. Dialogue, training and action planning are not, as I have already suggested, discrete activities, but often and usefully interrelated, at whatever level. Bipartisan dialogue may continue within the context of a multi-party, regional framework – in which case the needs of those who are not directly involved in the bipartisan dialogue need to receive attention in their own right – and attention may need to be given to additional conflict axes.

My fifth reflection is that one of the most important roles facilitators can play is to enable participants to trust their own capacity to deal with conflict, and to propose ways in which the most difficult and important issues can be addressed. Transparency on the facilitators' part will help ensure that participants are not plunged unwittingly into waters too deep for them, but have a voice in decisions about the process. At the same time, knowing that their facilitators trust their capacity for coping with the conflict will help them trust themselves.

My sixth reflection is that participants not only can, without detriment, know, but probably need to know, that their facilitators have values to which they are committed and their own feelings about the events and relationships at issue. What is essential is that they can be trusted to fulfil their role for the benefit of all concerned.

My seventh reflection is that funding agencies have the power of life and death for this kind of work. They can block or delay progress and, if organisations are not very clear and vigilant, distort their purposes. Equally, they can make the work possible and offer stimulus and undreamed-of opportunities. More dialogue is needed between them and those whose work they fund, so that the need for accountability can be married with the need for flexibility, and so that the needs for long-term funding can be met, as well as the need for readily accessible contingency funds. Working in the vicinity (both geographic and temporal) of war means dealing with all kinds of practical difficulties and unforeseen costs.

My eighth reflection is that there is a danger that, in addressing the major conflicts in a region, marginalised minorities (like the Hungarians living in Vojvodina, or the small minority groups in Kosovo/a, or the Roma people everywhere) are overlooked or excluded. Perhaps, in this case, a regional programme is needed for minority minorities. Equally, divisions which transcend all groups, like that of gender, may also need their own forum if justice is to be done to them.

My ninth reflection is based not only on events and experiences contained in this story, but on my broader experience of work in Kosovo/a, which has impressed on me the importance of external 'power players', the 'internationals', in regions of conflict. They may not be present in these workshops, but they cannot be ignored in reality. Maybe they should be brought in from the cold and included in these grass-roots deliberations – if only they would come. Local plans need to take into account their presence, actions and impact, coping with their influence where it is negative and harnessing it where it can help.

Recent work with ethnic Albanians in Kosovo/a has convinced me that 'internationals' of another kind (that is, external facilitators) can be useful in helping to create a space for local people to name the intimidation and silencing of dissident voices that goes on within their own community, and so to find the courage and make the strategies to confront them. At the same time, however, internationals can be treated as an audience for explanation, complaint and blame, diverting potential local actors from focusing on how to take up their own responsibility and power. Where there are local leaders and facilitators with the courage to speak out and liberate others, their work will be of paramount importance.

My tenth and final reflection is related to the value of work of this kind. Each of the workshops I have described cost thousands of pounds. How can we know that all this effort and expenditure is producing anything significant? Surely events have proved otherwise? How can working with small groups of young people make a difference? It was part of the institute's original aim to contribute, through its work with young people, to a climate in which a political solution could be found to the conflict over a territory's political status. The situation is very different now. Although the status of that territory has, *de facto*, been changed, it is still unresolved, and the region is fraught with acute local tensions and conflicts that need somehow to be transcended. There is an urgent need for education, economic input and development, and support for every kind of democratic endeavour. Despite and indeed because of this plethora of need, we should not, I believe, underestimate the importance of working with those people who are committed to the creation of a different kind of society and who have the energy to work for it. However small and few the seeds sown, they hold promise and can multiply – but they will need nurturing if they are not to die or be wasted.

9 Putting Respect into Practice

In my discussion of conflict transformation and culture (Chapter 2), I argued that the value of respect was universal and could provide a common, cross-cultural basis for conflict transformation. Respect is a recurrent theme in the workshop accounts that have formed the body of this section. In the journal extract quoted at the end of Chapter 5, I affirmed its importance both as a central concept for nonviolent approaches to conflict and as a reference point in dealing with cultural differences and dilemmas. I also proposed it as the 'litmus test' for what was acceptable in cross-cultural training. Before returning, in Part III, to the broader discussion of support for conflict transformation, I will summarise my thinking on some of the issues raised in Chapters 5–8 which I have found important in workshop facilitation.

It will be clear from my workshop accounts that the meanings of respect are many, and that they are affected by deep-seated assumptions about human society and relationships: different concepts and units of human collectivity; different ideas about the relationship between individuals and the communities of which they are part; and different constructions of gender and gender relations. The workshop accounts illustrate the tensions surrounding culture and difference when associated with asymmetrical power relations and historic and continuing injustice and the implications of those tensions for workshop facilitation. They raise questions about the role of outsiders in conflict, when those outsiders are seen as representing the globally dominant culture and are working in 'the majority world' (Alexander, 1996). They also, perhaps, suggest that such roles may still sometimes be useful, if uncomfortable.

RESPECT IN PLANNING

Respect for local needs and realities will be expressed in careful preparation. Workshop planning, if it has a regional or local focus, should be done by people with a close understanding of the region in question and of the conflict or conflicts with which participants are likely to be concerned (or which the workshop is supposed to help address). If the group which is designing the project is an outside one, then it will need the help and advice of trusted local contacts. Local knowledge should include cultural issues that have an impact on the conflict and which could be important in a workshop; also the power relationships that are important in society and need to be addressed if the conflict is to be transformed. This understanding will inform decisions about who will benefit from an opportunity to build their capacity to play a constructive role in the conflict or who should be included in dialogue. It will also have implications for workshop design, content and conduct.

Respect is further demonstrated by the choice of local facilitators or trainers, whenever they have the necessary skills and are likely to be acceptable to the group. (This may not be possible in dialogue workshops, or training workshops with participants from conflicting groups, where local facilitators may be regarded as partisan and facilitators from outside the region of conflict may be the only ones who are trusted.) Wherever they are from, facilitators will need to be well informed about the conflict(s) the workshop is to address, as well as the necessary cultural and political context. They need to be properly briefed and supported by those planning the workshop.

It is important that risks, both physical and psychological, are minimised. In some situations even participation in such an event might physically endanger participants. Depending on the context, it may be safer for participants to meet outside their own country, perhaps on neutral ground. Care for participants' psychological well-being is largely the responsibility of facilitators, so it is important they have the skill and experience to create a space and hold the boundaries for constructive interchange and to prepare participants for 're-entry' into their home context at the end of a workshop.

FACILITATION AND THE RESPECTFUL EXERCISE OF POWER

It will have become clear that to act respectfully and with integrity to all people in all circumstances is, to say the least, very difficult. Conflicts are part of the dynamics of group work in any context and, when cultural differences and political conflicts are added, can make for stormy episodes and be difficult to address. Although facilitators cannot be held responsible for all eventualities and the workshop itself needs to be well conceived, planned and prepared, their role is nonetheless crucial.

Although it is not the facilitators' job to dominate, it is their responsibility to exercise power in relation to many things. If the authority to do so is to be given to them by the participants, it is essential that they should command their respect. The proper and effective way to command respect is to give it, attentively and unremittingly, to the group as a whole and to each of its members. Facilitators' respect will be evidenced by consultation with the group about the things that affect it, in terms of both the process and the content of the workshop. Consultation will be needed at many points, not only at the beginning, when agreement is reached about the proposed agenda and ground rules, and at the end of each day, but whenever a significant decision needs to be made, taking note of changing energies and needs. Since groups are made up of individuals, consulting the group will entail weighing and co-ordinating differing views, drawing on the facilitators' own experience and explaining any decisions taken by the facilitators rather than by the group as a whole. Although the exercise of power on behalf of a group may be required of facilitators, manipulation is not. Respecting participants means being open with them about what is being done and why.

One responsibility of facilitators is to ensure that power is shared within the participant group so that all have a voice, reminding participants when necessary of the agreements they have made about listening to each other, respecting different views, sharing time and so on. They will be able to model respect for each one by the way they listen and respond to every contribution, not by adding many words of their own, but by brief marks of appreciation, by pausing for other responses, or by incorporating the ideas which have been voiced into visual presentations and into further work. If their respect for participants is genuine, valuing what they contribute will come naturally. If ideas are collected in a 'brainstorm' (that is, without comment in the first instance), then each should be recorded as given, not altered to suit the facilitators' own thinking, but different viewpoints can be discussed afterwards. Respect and agreement are not necessarily the same thing.

Power and knowledge are closely related. Facilitators of training workshops are expected by participants to bring with them a particular expertise. By adopting a broadly elicitive approach to training, facilitators respect the expertise of participants, enabling them to build on their existing experience and understanding. By offering them new concepts, models and processes, they respect their desire to acquire new knowledge, to be challenged and to have their thinking expanded. Paying careful attention to the mix of eliciting knowledge and making input, responding to participants' requests and at the same time respecting their own pedagogical viewpoint and values is one of the balancing acts facilitators have to perform.

Facilitators may need to act decisively at certain moments. Though they will remain accountable to the group through the evaluation process, it is important that they should respect their own power and responsibility to act for the group's well-being and to provide safety for each participant by 'holding the ring'. The greater the in-built tensions within the group, the more important this aspect of facilitation will be. Sometimes safety issues go beyond the psychological, and have implications for the physical security of participants after the workshop. Being clear about confidentiality agreements can be crucial. It can also be important to assist participants in weighing the risks associated with any commitments and undertakings they make during the workshop. However, respect also implies acknowledgement for their capacity to take responsibility for their own decisions.

Power-sharing between co-facilitators provides a model for co-operative approaches to power and support for dealing with demanding situations. When male and female or black and white facilitators work together as equals, the additional symbolism can also be important. If, for any reason, facilitators work together in unequal roles (for instance as 'lead' and assistant) their mutual respect and sense of underlying equality will be all the more important. It will be unfortunate if the functional inequality between them mirrors patterns of inequality elsewhere (for instance, in relation to ethnicity or gender).

RESPECT, POWER AND CULTURE

If the workshop's purpose and agenda have been agreed from the outset with organisers and participants, if the values on which it is based are clear and participants have been given some idea in advance of the kind of activities that will be involved, I believe that facilitators can be unapologetic about its general style. However, they should remain sensitive to the needs and responses of the group and of individual participants, not wilfully disregarding their feelings, but respecting their limits, and taking care to judge if and when it is appropriate to bring into the open any tensions which may be making themselves felt. Although it is difficult in some cultures to express negative feelings, it should be made clear to participants that in this context it is important that they speak up about anything which is proposed, or taking place, which is unacceptable to them. The facilitators should be aware in advance of any cultural sensitivities likely to impinge upon the workshop and any cultural assumptions which may shape participants' response to certain issues.

Cultural sensitivity and respect do not preclude challenging cultural patterns and norms. A workshop which was not at all challenging at this level could be regarded as a non-event. But the challenge should come through the provision of a framework for self-challenge, and facilitators should make it clear that they live in a critical relationship to the norms of their own culture. To become ever more aware of her or his own assumptions, cultural or otherwise, is therefore a crucial element in the development of a facilitator's own competence.

In addition to being self-aware, facilitators need to be aware of the contextual relationship between themselves and workshop participants and in particular the perceived power relations between them so that these are taken into account in the way they conduct themselves and frame proposed activities. It may on occasion be desirable to name (tentatively) the difficulty which this context and these possible perceptions present. What will always be most important, however, is the general demeanour of the facilitators – the respect they actually have for participants and the way that is made apparent.

CULTURE, GENDER AND HIERARCHY

Most participants, in most workshops, will come from hierarchical cultures. This applies in the West as well as elsewhere. Workshops of the kind under discussion create temporary, non-hierarchical communities, in which even the 'leaders' frame themselves as 'facilitators', that is, they present themselves as having an enabling function in a process in which others are the main players. Participants all have equal status within the process, leaving any external hierarchical relationships behind. Within the workshop context, all take their place and might be thought to appear as equals, but one distinction remains: that of gender. Since gender relations in all our mainstream cultures are, to some degree at least, unequal, it is not surprising when gender inequalities make themselves felt in different ways within workshops. Sometimes they are reflected in what participants say, at others in the way they behave, and at times by the predominance of men or women within the group. (If it is a workshop for people in positions of power, men are, despite the best endeavours of organisers, likely to predominate; if it is one for people working in certain spheres at a more 'ordinary' level, women may be in the majority.)

After much reflection, I have concluded that although I cannot, as a workshop facilitator, be responsible for the realities of the contexts in which participants have to act, I can and should encourage organisations for which I work to consider gender inclusiveness in workshop design and invitations. In the workshop itself, I can help participants to be aware of what is happening within the group and try to give equal space to all. I can also invite participants to elaborate on statements they make in relation to gender roles and related customs, providing the framework in which these issues, like others, can become a matter for reflection and evaluation. I can give encouragement to women within the group who wish to pursue questions of gender within the workshop and afterwards. I can also propose that separate workshops are organised to give women more space, to address women's experiences of conflict and violence and to explore what they can do to overcome violence against women and to work for cultural change. As a woman, I sometimes have difficulty in managing my own feelings in the face of certain statements and behaviours and need to process those feelings in a safe place afterwards.

CONTEXTUALISATION AND EXPERIENTIAL LEARNING

Whatever the conceptual elements of training or the skills to be developed, those concepts and skills will need to be 'contextualised', illustrated by, or applied to, particular situations – arguably the situations with which participants are most familiar or which are most relevant to them (see the Zimbabwe workshop account in Chapter 6). I believe there are several different ways of putting ideas in context, and that all have their uses. There is the context of participants' past and current experience, upon which they can draw and to which they can relate the ideas and exercises proposed. There are the contexts introduced by illustrations brought in by the facilitators that may come from situations close to the participants' experience, or from far away. One advantage of choosing examples from a context that is in some ways very different from the participants' own (so long as it resonates with their experience) is that they will be less preoccupied with the 'facts' of the case, and with the emotions which it engenders and better able to concentrate on the process or method of analysis in question. Contrasting differences can be as illuminating as noting similarities. Some participants also find it inspiring to hear about (or see videos of) people in very different circumstances who are nonetheless struggling with different versions of the same problems, using methods in ways which they can translate into their own situation. But that translation is vital. Whatever context is given in the first instance, if there is no application by participants of what they learn to their own contexts, the learning will be, to say the least, incomplete.

Although context can be provided by examples from outside the workshop, it is also given by the processes and dynamics of the workshop itself. As the accounts of the Geneva and Balaton workshops suggest, the learning which can be achieved in this way can be the most powerful of all. For that reason, dialogue workshops are an occasion for learning *about* conflict and ways to address it, as well as for dealing directly with it. But, as the Harare workshop demonstrated, to learn directly from the conflict experienced in the workshop itself may be very difficult, especially if the facilitator(s) are in some way implicated in it. It is also my experience that to try to learn from a conflict within the workshop while that conflict is still going on, or when feelings are still raw, is likely to prove counterproductive in terms both of learning and of conflict management. Theorising about intense feelings, at the time of their intensity, is usually perceived as disrespect for them. It is also likely to be seen as manipulative on the part of the facilitators, and as an abuse of their power. (I would *never* engineer a conflict, though I might tentatively name an underlying conflict I perceived as waiting to be addressed.) Reflection implies, for most of us, a degree of emotional distance. It takes a great deal of training to learn to be reflexive even in times of stress, and to take an external perspective while remaining present *in* a situation. These are indeed skills which facilitators need to develop and apply within workshops. The way in which they are able (or not) to facilitate the transformation of conflict within the workshop process will provide a model from which participants will learn, if not at the time, then upon reflection in quieter moments. On occasions when this is achieved within the workshop itself (and I find base group evaluation can help in this) I regard that as an important bonus.

THE CARE AND DEVELOPMENT OF FACILITATORS

Facilitating workshops is a demanding business. As already indicated, having a co-facilitator is the most important form of facilitator care.

Since comfort in co-facilitation presupposes compatibility, in terms of style and personality as well as values and concepts, care should be taken that such compatibility exists. It is good practice for co-facilitators who have no previous experience of working together to meet and discuss these issues before committing themselves to doing so. They will also need time for planning together. Compatibility need not always means similarity. Facilitators may have different strengths and expertise, but each needs at least to understand what the other is doing, so that s/he is able to offer support even in sessions in which s/he is not taking the lead.

Facilitators should not be responsible for logistical arrangements, but only for the actual facilitation, since their time will be more than occupied. It is difficult for them, once a workshop is under way, not to work around the clock. There are often pressures from participants to spend time socialising with them in the evenings or to attend optional sessions which they may organise for themselves. These extra-curricular activities are important and facilitators will wish to show that they hold them to be so. However, their first task is always to ensure not only that everything is in place for the next day, but that they themselves are rested and ready. They will need to hold a review of the day that has just passed, and a brief evaluation of it. (If base groups have reported on their own evaluation directly to the facilitators, and not in plenary, their reports will need to be digested and summarised.) Implications for the next day's agenda will need to be drawn, and the necessary adjustments made to existing plans.

If an open approach to planning has been taken, the next day's agenda will have to be designed or at least adjusted and the lead facilitation of different elements allocated to individual facilitators. The agenda will need to be written up, and any other flipcharts, printed materials and other equipment prepared. Even if social invitations from participants are refused – which is difficult – the work is unlikely to be finished before bedtime, and sleep may come slowly to the over-stimulated mind. I know of no solution to this problem, but bearing it in mind may encourage facilitators to protect themselves a little, by planning their evenings and being firm with themselves and others about their need for rest. I would also suggest that facilitators arrive at the workshop venue ahead of participants, so that they have a chance to rest after travelling, so that they do not start the workshop already tired. And I would argue that facilitators should have their own separate rooms for the duration of the workshop, even when participants are asked to share. This is a question not of status but of need. It is in everyone's interests that those who carry the facilitators' responsibility should be kept in good shape. Facilitators themselves should try to ensure that their work agreements provide for these things.

It is important for facilitators to keep developing their ideas about workshop content and methods and to be open to fresh thinking and techniques. Working with different co-facilitators helps me to break the mould of my own thinking and to see and do things in new ways. The most important source of new thinking is the challenge presented by the needs, perspectives and insights of each new set of participants. Necessity is the mother of invention. New situations demand new responses, in workshops as elsewhere in life.

Workshop facilitators have their own needs – for identity, credibility, approval and survival in their chosen field (see Rouhana, 1995). If these needs are not to get in the way of sound judgement about what is really needed 'on the ground' and who is best equipped to meet that need, self-awareness and self-questioning will be required. Both facilitators and the programme managers for whom they work will benefit from opportunities for 'peer supervision' – mutual reflection and support – in professional networks and seminars whose object is to provide space for learning, outside the sphere of competition and the self-justification which goes with it.

RESPECT IN THE FORM OF REALISM; ONGOING EVALUATION AND FOLLOW-UP

Respect requires honesty not only in the form of realistic goals for workshops but of realistic claims and assessments about their actual impact. Ongoing evaluation during the workshop process will help ensure its maximum relevance and utility for participants and participants' final evaluation will give some measure of this. What is harder is to assess is the subsequent impact of their new understandings and skills on what they are able to do and the effect of that on the situation in which they are actors. Sometimes there are clear outcomes in terms of new relationships, activities and institutions. Within their own sphere of influence, these may have clear significance or some known effect. When the participants have come from very disparate backgrounds and the primary purpose of the workshop has been to build personal capacities, rather than to have some immediate impact in a given situation, it is hard to measure the effect, except when it is possible to extract feedback from them at a later date – which in practice is often difficult.

A workshop is not a magic wand, capable of instant, do-it-all magic. It is a small contribution to educational processes that can support the transformation of attitudes and perceptions, resources and skills. Workshops work best when they are seen as part of ongoing programmes and relationships of co-operation for change. Resourceful participants will find ways of using what they gain, even from a one-off workshop, but organisers should have an eye for follow-up and be ready to help nurture the ideas and projects that may result, if not directly, by recommending others who can. The open-endedness of training is encapsulated in Lederach's idea of the 'permanent course' (Mennonite Conciliation Service, 1997), in which a series of training workshops continues as the regular provision of a time and place for colleagues to meet and develop their understanding and practice.

MONEY MATTERS

It is usual, when outside organisations arrange workshops in situations of conflict, for the costs of participation to be covered by them, mostly with the help of some funding agency or agencies. This is, in itself, at times a source of resentment for local organisations and participants who feel themselves to be put in the position of clients. The question of whether or not to provide *per diems* – daily financial allowances – for participants, so hotly contested in the Harare workshop, is one which has to be faced. One argument in favour of them is that, when participants have little money of their own, receiving a daily allowance in compensation for loss of income may be a precondition for their absence from work. Or *per diems* may be seen as 'pocket money' which enables participants to go out into the town or socialise in the ways that they might not otherwise be able to afford, especially when exchange rates are disadvantageous to them. One argument against giving money to participants is that they may regard workshops as a kind of paid holiday and come for that reason, rather than because they are seriously motivated to address conflict. It may also be felt that to give *per diems* is simply an inappropriate and unnecessary use of money, or that it is demeaning. I remain ambivalent and believe these decisions can be made only in context.

Part III

Looking to the Future

10　Good Practice

The focus of Part II was on workshops as the most widely practised form of nonviolent conflict intervention. In this third and final part of the book, as in Part I, the focus will again be broader, taking in the more general field of work in support of conflict transformation. This short chapter will be devoted to the question of 'good practice'. Once again the litmus test will be respect. Respect, in turn, implies the will, effort of judgement and application to be as effective as possible in the work undertaken.

If respect is the theme tune for this book, the accounts of practice presented in Part II demonstrate the minefield of complexity, cultural difference and unequal power that make it so difficult to walk to that tune. From this, one thing is clear to me: there can be no blueprints for *what* should be done – beyond the proscription of blueprints and the kind of top-down, generalising certainty they imply. Yet I would argue that it is possible to formulate some ideas about *how* things can be done, ideas that can point us in the right direction and inform our choices, plans and actions, ideas that can help us to avoid disrespect, irrelevance, waste and unnecessary risk and can increase the likelihood that we may be of use to those who are trying to overcome violence in their own context. As with so many fields of endeavour, it is often through observation or experience of what is badly done that it becomes clear what it would mean to do things well.

Here, then, under different headings, are some suggestions for the avoidance of bad practice and the application of respect in what is, inevitably, for both local and 'international' actors, risky and difficult work.

ATTITUDES, VALUES AND RELATIONSHIPS

Humility

There are several reasons why those who try to transform conflict should approach their work with humility, or, as Rouhana (1995) would call it, modesty. One is that those who work in this field are relatively few in number and small in power, so their efforts cannot stand alone but must rely on the cumulative effect, over time, of many different players. Another is that, in line with chaos and complexity theory, it is impossible to trace the effects of most things that are done beyond their most immediate outcomes. A third reason for humility is the smallness of our understanding and the limitations of our insights and capacities in the light of the complexity and size of the problems we would address. A fourth, for those of us who come from 'the West', is the knowledge that our own countries, directly or indirectly, are responsible for much of the world's injustice and instability (though of course not all), and that we have not been entirely successful in managing our own inter-ethnic relations or in banishing poverty and injustice from our own societies.

Humility is also needed about expertise. To be to any degree adequate as 'an expert' in conflict transformation is unlikely, given the intensity, longevity and deep-seatedness of many conflicts. Those of us who make this work our profession should respect the knowledge of those who live with violence of necessity and have learned from direct experience. We should not pretend to offer 'the solution' to what will take many lifetimes to change fundamentally.

Respect for Local People, Partnerships, Expertise

Those of us who work in regions outside our own should do all we can to rid ourselves of missionary or imperial attitudes and behaviour, understanding that while we may have the advantages of an external perspective and some useful expertise, we will at the same time lack the insights and connectedness of those who come from and live in the region in question, as well as the rights which they have, as citizens of their society, to determine its future. We will also have much to learn from them for our own wider understanding of the field in which we work. If we can bring helpful support, moral and practical, and some useful resources and connections, we may be accepted as partners by local actors and have something to contribute as well as to learn. We should do what we do in ways that maximise the development of local capacities.

Linguistic Sensitivity

The use of language and languages carries intended and unintended messages about attitudes to others and to power. The old colonial languages are often useful to all concerned in an exchange, but those who come from the countries where they originated should remember their continuing symbolism. In conflicts where language differences are symptomatic of deeper divisions and power relations, any choice of local language for interpretation will be difficult and will need to take these sensitivities into account. What is to be avoided in all instances is the use of language, consciously or unconsciously, as an instrument of domination – and if possible the perception that it is being so used.

The goal should be effective, nuanced communication which is open to all participants in a given process. It is important to remember that words represent mental constructs which do not always translate from one language to another. This can lead to enlightenment, but, overlooked, to misunderstanding and failure to recognise difference.

Respect for Different Perspectives, Cultures, Values

As I have suggested in the earlier parts of this book, respecting other cultural perspectives is not always a straightforward matter. Some of the norms of our own societies, as well as those of others, may contravene our most cherished values, in particular those underlying the 'project' of conflict transformation. Respect does not preclude disagreement. It is nonetheless incumbent on us to respect those who think and live differently from us, whoever we are, to recognise that other cultural customs and world views are as taken-for-granted as our own, and that we should not be blind to the fact that our opinions and ways of doing things are no more than that, representing our own culturally formed habits and ways of seeing things, and our personal gropings for pattern and meaning.

As far as is possible without violating or betraying our own deepest values, we should respect and adapt to other customs when we are guests, and be open to learning from them. We should also recognise that the perceptions of events and actions, and the meanings given to them, will be culturally influenced, and that we need to understand those interpretations.

Self-awareness and Transparency about Assumptions, Values and Goals

In order to give recognition to other cultural customs and perspectives, assumptions and values, it is necessary to be (as far as possible) conscious of one's own. This involves cultivating the habit of self-awareness and self-challenge. (Recently I met an important staff member in an equally important conflict-monitoring organisation and asked him what the values that underpinned their work were. With surprise in his voice he replied that his was an organisation whose work was at the political level and which did not, therefore, 'have values'!)

Sometimes organisational or personal interests outside our acknowledged values and goals are in fact influencing our actions and direction. Our awareness needs to include such influences (ambition, desire for recognition, financial pressures, for instance), so that motivations which are incompatible with our considered and acknowledged aims are prevented from distorting actions and relationships. For instance, we should be wary of personal or organisational 'expansionism' and consider carefully whether a proposed undertaking is something that will be really useful and, if so, whether we are the people best equipped to do it.

If we are clear with ourselves about the values which motivate us and about our own 'agenda', then it is possible for us to be open and honest about these things with those with whom we seek to build respectful partnerships.

Clear Plans and Agreements

Knowledge of and respect for each other's values, perspectives and goals will enable clear plans and agreements to be made together to which both (or all) partners are committed, naming the distribution or sharing of power and responsibility, and giving all concerned the opportunity to make informed decisions. Agreements and relationships may need to be changed, but that should also be done openly and by agreement.

For those who want to combat violence and injustice, wherever they live, the prospect of being paid for what they do can be a bonus. If this is their only work, it is usually a necessity. In economies where making a living is very difficult, or for people anywhere in difficult financial circumstances, having time for unpaid work may seem a luxury. At the same time, movements depend on volunteers. Most of those who work in them as professionals are likely to work unpaid as well, either by putting in extra time or in some capacity outside their professional duties. There needs to be a relationship of respect between paid workers and volunteers, and, again, clear agreement about roles and responsibilities.

POWER AWARENESS AND ATTENTION TO JUSTICE

Programme Design

In order to make well-informed decisions for effective action in support of conflict transformation, it is necessary to pay attention to the role played by power asymmetries and to the ways in which these are being exploited and experienced by the different parties. The unavoidable power of Western members of partnerships places the initial responsibility (and need) for such awareness on them. They should start from local perspectives, paying great attention to listening and constantly opening the space for input to be made and disagreement to be voiced.

Working to Include Those who are Marginalised

Although it is relatively easy to assert the importance of including in conflict transformation all those who are affected by the conflict, it is hard, in practice, to be effective in changing existing realities without being influenced by them. Wanting to work with those who currently have most influence implies not working with those who are excluded from power. To take a clear example: the Roma populations in many areas of conflict suffer greatly from their social, political and economic exclusion, but have little or no chance of changing the wider political conflicts that affect them at least as much as other groups. Although this dilemma needs to be addressed in the light of specific situations and goals, I would suggest that for those who see justice as an aspect of peace it is essential not to perpetuate exclusion but to try as far as possible to allow all voices to be heard.

There may be a conflict between commitment to respecting local perspectives and commitment to this kind of inclusion. (Witness the initial lack of enthusiasm encountered in Kosovo/a for even thinking about Roma.) In such cases it will be necessary to find ways of resolving or living with this dilemma.

In the meantime, separate work is needed for the empowerment of excluded groups, so that they become more effective in advocating their own needs, dignity and potential to contribute.

Attention to Gender Relationships and Dynamics

Women, though numerically the majority of all populations where female abortion and infanticide are not used, everywhere constitute a 'power minority' and in all societies are either excluded from public positions of power or under-represented in them. The problems involved in their inclusion are similar to those related to other marginalised groups. However, if the use of gender as an instrument of domination is to be addressed (contributing, as it does, to the culture of violence) and if the skills and insights of women are to be fully engaged in the service of conflict transformation, it is vital that explicit attention be paid to the inclusion of women in projects and programmes designed to support it. It is also vital in analysis to examine the ways in which women are affected by and can have an impact on a given conflict.

As with other marginalised groups, it is useful to consider the possibilities for contributing specifically to the empowerment of women and to their potential as peace-makers at every level.

RESPONSIBLE ENGAGEMENT; CARE AND QUALITY OF SERVICE

Respect for partners and intended 'beneficiaries', not to mention those who fund our work, must include care and responsibility at all levels for the way that work is done.

Assessment of Organisational/Personal Capacity and Resources

Before a commitment is made to a given project or programme of work, whether by organisations or individuals, they should be sure they have the necessary local knowledge and connections, and the skills, capacity (in terms of time, personnel and infrastructure), and the likelihood of access to funding. They should not raise unrealistic expectations or promise what they cannot deliver.

Initial Analysis and Planning

Proposals and agreements should be made on the basis of sound local information and careful analysis and planning, including clarity of purpose, chosen methods and their rationale, partners, entry points and participants, resources, potential obstacles and how they are to be addressed, different roles and responsibilities, and provision for monitoring, evaluation and potential modification.

Weighing Risk

In project or programme design, the 'Do No Harm' principle should be applied (Anderson, 1996). Organisers should seriously consider whether there is a significant chance that planned activity could contribute to a worsening of relationships, rather than their improvement (for instance, by seeming to favour one side or by being too ambitious about the kind of encounter that is politically and emotionally manageable at a given time).

It may be impossible to work in tense and violent contexts without some risk – to staff and volunteers, local and international, to those who choose to be involved in the initiative, or to the reputation of the organisations concerned. However, it is important that all concerned are as far as possible aware of the risks, and incur them knowingly. (Even to be approached may in some circumstances endanger someone. In such cases, contact should not be made without good reason to believe that the person concerned would welcome it.) Plans should be made for monitoring and managing any risks that have been identified.

When the risk taken is one of losing face or reputation through failure, it may be that the needs of the situation will be judged to make such a risk worthwhile – although that too will need to be weighed as it may jeopardise future work.

Confidentiality

In sensitive situations, if trust is to be possible, it is essential to respect the reputation and safety of those with whom one works by honouring the confidentiality of what is said in private, whether the context is that of a private conversation, a workshop or any other meeting. It is also vital, in some circumstances, not to make public the very fact of some meeting or behind-the-scenes communication.

Attention to Detail

The chapters in Part II about workshops demonstrate the importance of attention to detail in both their conceptualisation and their implementation. Though the things needing attention will differ with the work and its circumstances, carelessness in organisation, venue, decisions about whom to include and exclude, provision for interpretation, timing and communication can lead to wasted efforts and even to unnecessary risk.

Ongoing Analysis and Flexibility in Implementation

As implied above, analysis must be ongoing and plans may need to be modified in the light of experience and in response to what is revealed by monitoring and evaluation. Furthermore, what made sense when plans were laid may make no sense in the light of major political changes, local reversals or unexpected outcomes. Respect implies a serious intention to be maximally effective in the service of identified goals, which may involve quite substantial changes to a programme as events unfold. If partnerships are genuine, local partners who are the main implementers of a programme may challenge their international counterparts to become involved in supporting work they had not envisaged. (For instance, one organisation working with a pro-democracy organisation in Fiji chose to support its pro-democracy partners in their decision to move from the role of broker of dialogue to legal challenger of government – a major shift which involved their international partners, as well as themselves, in risks they had not originally envisaged.)

EVALUATION

Challenges and Limitations

Although the importance of evaluation is generally accepted (given that conflict transformation initiatives are often quite costly and compete for funds with other important forms of assistance), it is often difficult in practice to predict or prove the impact of a given activity. While it is relatively easy to demonstrate that so many people were visited or involved, or that certain actions were carried out or topics discussed, or that so many people awarded so many marks out of ten to the facilitators, it is not usually so easy to show what effect all this had on those involved, let alone any consequent effect on the wider situation. Although sometimes such effects can be traced, and attempts should be made to do so, in the interplay of different factors and the complexity of influences, it may well be impossible to attribute particular changes to single causes. The impossibility of measuring impact should not, however, deter us from undertaking work that seems important. The more fundamental and long term the problem addressed, the less likely it is that effects of changes brought about by a given activity or intervention will be able to be measured.

Even when an intervention with a rather immediate purpose fails (for instance, when the objective was to bring leaders of conflicting parties 'to the table'), it may have had some effect which bears fruit at a later stage. Moreover, to make an attempt to achieve something, to go about it in a well-judged manner and with great care, on the basis of thorough information and sound analysis, using people with appropriate experience and skills, arguably deserves positive evaluation, even when it fails to achieve what it set out to do. 'Nothing ventured, nothing gained.' How many failed attempts have there been, for instance, to address 'the Middle East conflict'? Does their failure mean they were all ill-judged or badly done?

I ask these questions not to suggest that we should dispense with evaluation, but to suggest that it cannot always be based on provable impact or even on provable success or failure. What, then, is possible?

Assessment of Implementation

It is possible to identify the ways in which the implementation of an event or activity, or programme of activities, corresponded to their design. Did the practical arrangements work out well? Was the concept of the event validated in the way it turned out? Was the content of discussions in line with that aimed for, were interactions favourable, did those involved make the hoped-for commitments for the future? As far as it is possible to judge, do the immediate apparent effects and participant evaluation give grounds for hope that they will contribute to the desired process or outcome? What contributed negatively and positively to the way things went? Were the skills of those involved, for instance, facilitators, advocates or mediators, adequate to the task? How could they have been improved or at what points could things have been done differently? If things went wrong or were disappointing, could that have been avoided, was there an avoidable weakness in the concept or design, or was it a well judged and implemented attempt at something that was worth trying and can be learned from?

Evaluation can be seen as being of two kinds, with different purposes: 'summative' and 'formative'. The former is intended to hold those who gain support for a particular endeavour accountable for working responsibly in its design and implementation. It involves a review of the quality of their work and an assessment of whether they achieved all they could reasonably have been expected to do, in an unpredictable field and in the light of known risks. The purpose of the latter is to learn for the future, both about how things should be done in future in the specific situation in question and about effective practice more generally. Both forms are important, for different reasons and to different people; but in the development of good practice, the latter is paramount.

Assessment of Future Options

Evaluation can be formative of future strategy – *what* can be done – as well as suggesting *how* activities are carried out. What does it suggest about further activities already planned? Are they exactly what is needed or do they need modification? Should there be a change in direction or emphasis? Are there new initiatives to be pursued or new leads to be followed? Is it necessary to start again somewhere else?

Monitoring Outcomes

Sometimes it is possible to track the effects of particular initiatives, either on participants (according to their own assessment or as evidenced by, for instance, some new activity on their part) or, where the impact is immediate and limited, on events. Sometimes an immediate impact of that kind can be seen as potentially important in leading to the achievement of a more distant goal, but it is impossible to predict whether its ongoing effects will in fact reach that goal. In Moldova, for instance, where part of the country has seceded, NGO activity can be seen as contributing to keeping alive the political hope and will to avoid violence and to achieve a peaceful settlement; but that settlement remains elusive. Nevertheless, it would seem, on balance, to be worthwhile to continue the NGO efforts and associated expenditure.

LONG-TERM COMMITMENT WHERE NEEDED

In certain circumstances short-term action or support may be just what is needed: a particular piece or course of training, the development of an organisation, establishment of a communication mechanism or some other type of capacity-building; or a one-off catalytic intervention requested by conflict parties, such as the provision of a timely opportunity or urgently needed pretext for meeting. However, as the above discussion of evaluation would suggest, most activities and partnerships, if they are to be useful, imply long-term commitment. An endeavour, once begun, should ideally continue as long as the necessary conditions are in place and it has a chance of being effective.

Equally, this implies ongoing assessment of the relative utility of a given initiative or partnership and willingness to recognise when something has failed beyond retrieval or reached the end of its usefulness, or when changes in external circumstances have rendered it irrelevant. Partnerships may also no longer be appropriate if a local organisation is strong enough to do without external support. International workers should not be driven by a need to be needed.

RESPECT, POWER AND MONEY

As I suggested in Chapter 9 in relation to my discussion of *per diems*, Western organisations often (though by no means always) are the partners with access to money. This can give them, potentially, undue power in a given partnership, which they should take care not to exploit.

Skewed global power relations are also reflected in very different rates of pay. The purchasing power of a given currency in its own country is often very different from its purchasing power elsewhere. The average standard of living also varies greatly from one country or region to another. It is already problematic that a large proportion of funds raised in the West for use elsewhere is used in fees for Western consultants. The level of those fees, while it may (and I would argue should) be modest in terms of Western norms, is still very high compared with comparable fees paid in the places where they work. To pay locals at Western rates would be to set them apart from those around them, disturb the local financial ecology and attract into the field people whose overriding motivation is financial. To pay Western consultants at local rates when their main living costs are in the West would make it well-nigh impossible for Western consultants to do this kind of work except as a kind of hobby. Yet to pay local and 'international' colleagues vastly different amounts is invidious. These realities are uncomfortable, divisive and potentially poisonous for relationships, both interpersonal and inter-organisational. Perhaps the best that can be done is to talk about them honestly and work for a mutually tolerable and transparent formula, made on the basis of agreed principles.

PERSONAL AND ORGANISATIONAL INTEGRITY: 'PRACTISE WHAT YOU PREACH'

To have integrity and win the respect of others, individual practitioners and organisations in the field of conflict transformation should do their best to live out their own values and put their theory into practice, especially in the way they handle power relations, deal with conflict, respect and include different cultures and backgrounds and promote gender equality.

The organisation's culture should be in line with its collective vision, its articulation of its mission and goals, its policies and publicity. Its structures and practices should embody co-operative forms of power, being designed in terms of responsibility and function rather than hierarchy and involving collective processes for learning and the preparation of decisions. This will require transparency and clarity about the goals and realities of procedure and structure, with codes of practice to set standards and provide a reference point.

The good example set by such organisations can have a multiplying effect. By the same token, contradictory behaviour will undermine an organisation's reputation and influence. The sad irony is that it is often organisations with the highest ideals that are most torn by strife and confusion. High principles need to be coupled with business-like attention to systems and their development, and regular opportunities for review, including all staff and board members.

SUPPORT AND DEVELOPMENT

Working with violent conflict is stressful and can be both distressing and depressing. It is important that those so engaged should have the opportunity to talk about their experiences and feelings, rather than being left to deal with them alone. It is also important that they should reflect with others on their practice so that they can learn from successes and failures, developing their skill and understanding. Sizeable organisations should provide ongoing training for their staff, or at least occasions for collective reflection in which staff can encourage each other in strengthening personal and organisational self-awareness and maintaining the will to confront and wrestle with dilemmas and contradictions. In programme work they should build in processes for regular feedback, monitoring and reflection and consequent evaluation and adjustment.

Practitioners who work freelance may find support in organisations that employ them, as well as from individual colleagues. They may also participate in groups and networks for mutual support and the review and improvement of practice.

CO-OPERATION IN PRACTICE AND LEARNING

Organisations working in the same field can choose to regard each other as competitors or as colleagues with whom they can co-operate and learn. If their primary motivation is the well-being of those they serve, I would suggest that the latter approach is the more appropriate. Often several different organisations, local and international, are working to address the political conflict and its manifestations and causes in a given region. Although there may well be room for all their efforts, it can increase their effectiveness if they are intentionally complementary and the exchange of information and contacts can help all of them.

Several networks for learning and co-operation exist. I belong to one such professional grouping through which member organisations share information and experience, exchange insights and develop ideas. Its members also arrange seminars with wider participation and input, so opening themselves to challenge and stimulus from circles beyond their own.

I would argue that to contribute to the development of one's field is one aspect of good practice. It is essential that the body of thinking related to this field should be constantly reviewed and augmented, and that theory should be formed and re-formed in the light of practice. The development of theory is in turn related to the development of policy in political and funding circles and so can influence the context within which this work is carried out.

The theory has a long way to go, not only because the field is both relatively new and infinitely complex, but because it has to date been formulated, predominantly, by Western men, and needs to be challenged and rounded by the world views, insights and experiences of women and of theoreticians and practitioners from different parts of the world.

The scope of research should include not only what leads to and perpetuates hot conflict and what helps end it and prevent its recurrence, but it should also include the study of people's movements for change and what makes them effective.

GOOD PRACTICE AND FUNDING

None of the ideas outlined above can be put into practice without funding, and the way grants are currently made creates substantial obstacles for good practice. Procedures are slow, which in fast-changing situations can mean that interventions are delayed too long to be relevant. When grants are made piecemeal for individual projects, rather than for longer programmes, it is very difficult to maintain continuity or to retain the necessary staff. Moreover, the kind of sound organisational processes and structures discussed above and the vital space for ongoing reflection involved in the maintenance of good practice and development of knowledge, call for a level of organisational funding which is extremely difficult to maintain. Organisational resources are wasted on constant and inordinately time-consuming fundraising efforts and programmes and organisational capacity is lost because funds for continuation cannot be found. It would help a great deal if funding agencies would take into account the importance of sound and stable organisations as a basis for conflict transformation work, of continuity in programmes, and of streamlined grant application processes with quick responses.

While careful rationale and planning, accountability and attention to cost-effectiveness are vital, it is also necessary in this field to accept the unpredictability of circumstances and outcomes and the difficulty in demonstrating impact. There needs to be an ongoing dialogue between funding and implementing agencies about how the sometimes conflicting needs of each can be most effectively managed and rigour and flexibility combined.

11 Making a Difference: Challenge and Change

In this final chapter I will present the future of conflict transformation in terms of a series of challenges to existing cultural, theoretical and political norms and concepts. In doing so, I shall return to issues raised in my first, introductory, chapter and elsewhere: the cultural 'normalisation' of violence, oppressive models of power, and the impact of these on North–South and male–female relationships. Acknowledging the difficulties involved in transforming the culture and politics of power and conflict, I will uphold as fundamental the role of 'ordinary people' in the work of transformation.

CULTURE

Challenging Violence

If the project of conflict transformation is to be promoted, it will involve challenging the culture of violence that runs as a thread through most if not all cultures, whether more or less explicitly or strongly. In many societies, violent behaviour by civilians is outlawed. It is seen as unacceptable that in day-to-day life people should assault or kill one another. Even then, violence between young males is often 'normal'. And violence on the part of states and other collectivities is seen as their prerogative, and is prepared for through the training and maintenance of armies, with weaponry ranging from machetes to nuclear missiles. The history of a people's or a country's military exploits is usually regarded as a matter of pride, in a way that a personal history of violence would not be in 'modern' societies. If those of us who live in such societies wish to mount a challenge to violence as a means of addressing conflict, that challenge must include state violence, especially that done by our own countries, for which we share collective responsibility. We must campaign for all states to be held accountable for acts of violence – even those whose power currently defies challenge. Otherwise it will be apparent that economic power, achieved through past violence, puts those who hold it above the law and above morality, and our advocacy for nonviolent approaches to conflict will rightly be viewed with scepticism.

Challenging Constructions of Power and Gender

The culture of violence is based on the construction of power as the ability to dominate and control (with violence as its instrument). This way of understanding power is in turn connected to prevalent constructions of gender, specifically of manhood. 'Softer' models of power (often constructed as female) are central to conflict transformation. They spring from impulses of care (respecting humanity, responding to human need) and are associated with the notion or intuition of interdependence. As the world continues to shrink, the good sense of this understanding of power will, it is to be hoped, be increasingly recognised. The fortress and army mentality cannot, in the end, answer the needs even of the most powerful. More co-operative models must prevail.

If this cultural shift is to be achieved, 'mainstreaming gender' in all spheres of life must mean more than simply men 'including women' or even women taking their place at all levels (Reimann, 2001). It will mean the deconstruction of gender and the development of new ways of understanding and representing humanity, in all its variety. It will involve the establishment of the values of care, co-operation and respect as pre-eminent.

THEORY AND CONCEPTS

Developing More Rounded Theory

Although a substantial literature already exists and a great number of academics and practitioners are engaged in wrestling with the complexities of conflict and ways of addressing it, those complexities are so great, and the realities with which the field is concerned so interwoven with other realities, that far more needs to be done. I would argue that more theoretical input is needed from those 'on the ground'. Good theory is useful for them. Not only can theory illuminate conflict and suggest ways in which it can be addressed, it can give to those directly engaged in working with conflict a clearer understanding of their own part in events, enabling them to increase their impact. At the same time, they have a contribution to make to theory, particularly to theory about the kind of power and reality of which they have first-hand experience.

Workshop participants, for instance, are developers of theory, not simply its recipients, and workshop facilitators have excellent opportunities for learning about conflict and its transformation from the experiences and insights of those with whom they work – as well as from other practitioners and academics. That living knowledge, constantly formed and re-formed in interaction and in the light of experience, can bring more life and substance to theory about the role of 'people power' and the obstacles to it. And in cross-cultural work, it is the living process of theorising that provides a safeguard against cultural blindness and imperialism, revealing the possibilities of new answers to old questions and changing the questions themselves. It would be wonderful to have books written on the basis of symposiums, or collected interviews involving both grass-roots activists and political leaders in reflection on the work they do for peace.

For the rounding out of theory, more minds and voices from the two-thirds world and from women are needed – otherwise it will remain woefully incomplete and distorted. A genuinely inclusive engagement and debate would, no doubt, change both the emphasis and content but also the character of the content of the theory. For such a productive ferment to happen, it will be necessary for those within the field's academic establishment and profession to remove the invisible walls that cut them off from the daily prac-titioners working at the local level (I suspect far more women than men), meeting them where they are and inviting them into their learned circles.

In the South and East of the world, there is considerable interest in the theory and practice of conflict transformation and a great deal of relevant experience and knowledge, but there is much suspicion, too. In both theory and practice, there needs to be follow-through on the assertion regularly made in the conflict transformation field that conflict in itself can be both necessary and constructive. The splicing of nonviolence theory (born in India) with Western-born 'conflict resolution' is one way in which the political perspective of those who need conflict can be aligned with the viewpoint of those whose main preoccupation is with preventing or ending it. It is a way of marrying peace with justice by providing an alternative to organised violence as an instrument of change. If such a broadening and shift of emphasis were to take place within the field, the theor-etical agenda would change. From this perspective, for instance, more research is needed into movements, how they work and what makes them effective or blocks their success. This research would be relevant also for those whose need is to end violent conflict and who wish to develop peace constituencies or movements for peace.

The perspective of change and constructive engagement with conflict (rather than its prevention) is linked to an energy that flows against that of control. That energy could free the way for more exciting, exploratory theory, less preoccupied with pinning things down and making outcomes predictable (which they never can be) and more concerned with entry points for creative action.

Socio-political Levels: Spheres of Influence and Action

Judith Large, writing in Spring 2001 for the Committee for Conflict Transformation Support, describes the situation on the Molukan island of Ambon, Indonesia, while international intervention was under way but had not brought control:

> Ambon City is in total rubble and firmly divided between Christian and Muslim quarters. But people are talking, about jobs, about rebuilding, about how to govern themselves. Indonesian mediators have held first meetings with the police, a mixed force who had totally splintered during the fighting. The governor struggles to maintain a neutral zone around his office. There is talk of a mixed market. Gang leaders speak openly about their wish for jobs.

She goes on to say, 'This is not presented as a success story. This is a transition in progress largely from below.'

Even during wars individual actors, and indeed whole communities within warring territories, go on making choices, whether active or passive, about how to behave to each other and how to respond to events, creating a variety of realities on the ground, where non-war life is still lived. Those who attempted to avert a war may continue to act to uphold the values that motivated them, and make a difference to the lives and actions of many people, even while it continues – for instance, preventing or reducing political murders or evictions in a particular locality, working with refugees to help them establish micro-societies that uphold human dignity, or introducing new kinds of thinking and practice into children's education. They may even manage to maintain inter-communal respect and co-operation within their own locality, thus denying war its supremacy, surviving as small circles of peace that could not be broken, growing points for a different kind of future when at last it begins to take shape. Top-level decision-makers may be able to declare wars and sign treaties, but only people can build peace.

While the notion of different levels of society (depicted by Lederach, 1994) has made an important contribution to thinking about the powers and limitations of different activities and groups, it should be regarded as an entry point and stimulus for further thinking.

For instance, complex societies contain many power hierarchies, not just one, and although the decision to settle a conflict may be, apparently, taken by one or two human beings, wars often have multiple rather than single engines. Increasingly we are coming to recognise that the political powers of states do not control economic forces, whether in the legal form of multinational corporations or as exercised by 'unofficial' economic systems, including mafias. And those economic forces that are beyond the control of governments also have their own gunmen, militias or private armies to defend their 'territory' or ensure their dominance. We need to develop models that demonstrate different sources of power and their inter-action and potential influence.

The idea of leaders as a small and lofty group can tend to obscure their humanity and vulnerability, hiding the fact that even those 'at the top' are often motivated by personal ambition or fear of eclipse, disgrace or death, sometimes even by care for 'their' people (Curle, 1986). Their humanity makes them changeable and often unpre-dictable. For good and ill, their actions do not follow some inexorable, structural logic. Furthermore, leaders do not really inhabit a lofty, sealed-off eyrie. They have power only insofar as they can hold on (by whatever means) to the support – passive or active – of enough ordinary people. Wars usually need foot soldiers. (One surprising and grimly comical turn taken in recent history has been the new unwillingness on the part of the big powers to 'sacrifice' the lives of their own soldiers in war. The technological solutions to this dilemma make war, more than ever, a blunt instrument, to be used only when the power of the user is overwhelming and the enemy a sitting target.) More thinking is needed about the power and the vul-nerability of leaders in conflict, about the relationship between their human being (being human) and their function within the social structure. (This in turn connects with questions about the very notion of social 'structures', as distinct from the people involved in them.)

The distinction between grass-roots and middle-level actors is in many societies not real and certainly not fixed. Circumstances are too new and relationships too fluid. Political and social upheavals and displacement make for additional mobility, and young people, particularly, may 'move up'. Top-down politics is often part of the problem – often one of the causes of conflict. Arguably, it is the civil counterpart of war and bottom-up action is needed to challenge and change it.

The point of Lederach's model is to clarify spheres of influence, not to define limits but to suggest the importance of transcending them, and his argument is that actors from the middle level can have an impact on the levels above and below. The dynamic connections or continuities between these putative levels need to be further considered and mapped. And Lederach's concept (like that of Ropers (1995), about the realm of states and the realm of societies), could become the launching point for the creation of new 'models' of the different circles of influence in societies. It could unlock fresh thinking, not only on potential spheres of action for conflict trans-formation, but on the ways that one sphere intersects with or can influence another. In what ways do the different spheres of power at all levels interact and where are the points of leverage within and between them? In whatever sphere, in what ways can those at the community level influence 'higher' levels of activity and leadership? Could community organisations usefully extend their attention to questions of policy and the political context, without jeopardising their cross-community identity and acceptance? How can 'top-level' decision-makers involve other strata of society in moves towards the negotiated settlement of conflicts? Could intra-party dialogue forums between different levels and factions help to prepare the ground for political agreement between conflicting parties?

The development of theory in these areas would facilitate explor-ation and inform planning within organisations working for conflict transformation, and indeed within workshops, suggesting new locations for action within overall strategies, or even radically influ-encing the strategies themselves. (Some of the experience that could help in the formation of such theory already exists in the excellent series by Conciliation Resources, *Accord*, each issue of which charts processes in the search for the political settlement of a different conflict.)

It is easy to think that what matters most is what is done by those 'in power', and that those in power are those 'at the top'. When one thinks of wars like that in Sierra Leone, where it is clear that the vast proportion of 'ordinary people' have had their wishes and efforts for peace constantly rolled over by those who wished to perpetuate the violence, it is hard not to make this assumption. But closer examination reveals that the fighting has kept re-erupting not only because of the ambitions of leaders but because of the disaffection and hopelessness of many young men (and children) who feel themselves to be marginalised and without any identity or future outside their role as warriors. In order to prevent war or bring it to an end, it is necessary to win the commitment of actors at all levels to the agreements necessary for coexistence, just as overcoming terrorism will involve overcoming the alienation and anger from which it springs. Peace, like war, needs a 'constituency', and workshops of the sort described in Part II of this book may support the building of peace constituencies.

In the light of the apparent ease with which the kind of visions generated in workshops and the actions undertaken by citizens at the local level can be blown apart by decisions made at a level beyond their reach, it is easy to feel disheartened and to question the relevance of such apparently puny efforts. But if the goal of a course of action is to contribute to efforts to uphold human decency, in whatever sphere, then it is of itself significant. Even apparently futile action has moral and symbolic meaning that will remain as a reference point when the time to rebuild eventually comes, and will inspire others. (Remember the man in Tiananman Square, standing in front of the tank?)

Speaking Up for Mercy

I have commented elsewhere that in the mixed workshops where I have used the Lederach exercise (described in Chapter 4) to weigh together the four values of peace, justice, mercy and truth, the value of mercy seems to attract the least support. This has troubled me a little, since the human condition seems to call for mercy more than anything. It has been argued (Gilligan, 1982) that female moral development is centred on an integrative approach to needs and rights and the impulse of care, as against a discriminating and judging approach and the value of justice. Whatever the merits of such gender-based assertions, I believe that the value and concept of care, combined with attention to inclusiveness, can be important in softening the impact of the pursuit of justice. Perhaps if women took their place and were recognised as equal players in the field of conflict transformation, the impulse, and perhaps the concept, of care would take on a more important role. It is close in meaning to 'distributive justice', but has a gentler feel, maybe more readily compatible than justice with 'power for' as against 'power over', and its elevation would bring in its wake a greater acknowledgement of our common and communal need for mercy.

Recognising the role of mercy in tempering discriminatory justice with inclusive care might reduce the apparent tension between truth and justice on the one hand and peace on the other.

Re-working the Idea of Community

Those who are working for peace at the local level are often casting off the communal straight jackets which have been foisted on them and creating community across communal boundaries. The word 'community', always tenuous in its meanings, has been so widely abused, so frequently used to incarcerate and to exclude and to mask divisions and differences, that it is sorely in need of re-creation or re-appropriation. As wars and economic pressures scatter peoples who have shared the same geographical and social space and as, at the same time, through modern technology, new links become possible, it is a time to create new versions of community, new forms of belonging – supportive and purposeful 'families' or groups, whether local, regional or international. While physical support and companionships may need to be locally based, psychological, ideo-logical and moral support need not be limited by space. Global movements and alliances, while problem-laden, are still excitingly possible. At the same time, new forms of local organisational units for human society are needed that transcend the relatively recent notion of the state, which is proving so enticing and so divisive as the symbolic summit for collective identity.

Processes and Outcomes

In Chapter 10 I discussed the difficulty of basing the evaluation of conflict transformation on identifiable outcomes. The emphasis on outcomes, as somehow distinct from goals (a gift, I imagine, from the world of management: the NGO equivalent of 'product') is, I believe, both philosophically and practically unhelpful. Quite apart from the practical problems associated with trying to track or ascribe outcomes, I believe a more fundamental challenge is needed to the notion that conflict transformation is something that can or should be judged by them. Murder is not judged by outcome, except in the sense that actions resulting in death are seen as wrong. The idea that outcomes or ends can be separated from processes or means is in itself fallacious. It underlies the belief that violence can produce peace, when in fact it erodes the ground on which peace can be built, throws ever further into the future the occasion for breaking the cycle of violence and starting something new. Doing things constructively – managing relationships, respecting others, building bridges, improving institutions – *is* peace. There is no static, ideal outcome that can be arrived at once and for all: only people doing things, working at living together.

POLITICAL RESPONSIBILITY

Living the Future in the Present; the Seamless Garment

There can be no denying the difficulties and dilemmas inherent in shifting from one system and paradigm to another: of acting out respect in societies and situations where respect is denied or not yet embodied; of living a peace culture in the midst of a war culture; of acting nonviolently while violence is all around. Is it possible in circumstances where women are disempowered for women to be included in all peace-making processes? The immediate and urgent needs of now often seem to be in conflict with the more fundamental need to live the values that make for peace. For instance, does it make sense to try and involve women in peace processes, since it is men who have the power to deliver it? Is it possible for internal democracy to be seen as a priority by a group that needs to maximise its unity in the face of an external threat? Does it not make more sense to keep internal divisions buried till justice has been won from the enemy?

I have a Palestinian woman friend who was prepared to put gender justice and democracy on hold till a free Palestinian state had been established. She came to realise, however, that the cost, in terms of ongoing exclusion from power and the consequent exclusion of many groups from the 'peace process', made this 'patience' illjudged. She told me that she had concluded that internal and external peace and justice were inseparable, and that they needed to be understood as different aspects the same process, rather than as a series of discrete (if hypothetical) outcomes (cf. Dworkin's (2000) feminist analysis on the State of Israel).

This example also demonstrates the inseparability of peace from participatory politics. The field of conflict transformation, while it needs to maintain a distinctive focus, must recognise that it is embedded in a mesh of other projects or fields of endeavour – economic justice, political freedom, minority rights, racial equality and environmental protection, to name but a few – what a colleague of mine once described as the 'seamless garment' of different, mutually supportive and dependent movements for respect in human life.

Conflict Transformation, Development and Global Participation

The building of peace needs commitment at every level. Peaceful politics are possible only with willing consent and active engagement from all groups within society. They require the formation of structures, the development of systems and the ongoing activity of people of all kinds, and those people need the vision, skills and cohesion of thought that make these activities and development possible. Since the maintenance of peaceful social and political relationships involves the constructive, nonviolent handling of the conflicts that will inevitably arise, developing or maintaining the understanding and skills for this is a task for societies at any stage. Workshops can make a valuable contribution to the development of all these capacities and to the rebuilding of relationships.

Conflict transformation constitutes a developmental approach to peace-making and peace-building. Both development and conflict transformation put 'ordinary people' at the centre of their vision. In recent years, the connection between peace and development has been increasingly visible in the thinking and strategies of development agencies. Just as it is clear to those who work for conflict transformation that structural injustice is an obstacle to peace, so those who work for equitable development are faced with the fact that war prevents and destroys it. Many development agencies now offer training to their staff in understanding the impact conflict can have on their work, and vice versa, and to help them to incorporate into their work strategies and activities for addressing conflict (see, for instance, Anderson, 1996; and the website of CODEP – the UK Network on Conflict, Development and Peace).

Furthermore, the values of self-determination and consent grow ever stronger, at least in the West, running counter to the military imposition of 'solutions' to local conflicts. Within this framework, it becomes clear that those who should determine the outcomes of conflict and build peace are those with the most direct stake in it – the people who live in the territory in question – and that the only legitimate role for outsiders is in helping to provide a space in which local forces for peace and social organisation can assume responsibility. This is the theory of conflict transformation, and the credibility of those who wish to promote it abroad, in workshops and in other ways, will depend on their being seen to promote it at home.

Facing Our Own Responsibility

I believe that those of us who work internationally for conflict transformation have a fundamental obligation to do so in our own countries, living, not just promoting, democracy, doing our best to 'put own house in order'. If we have any regard for our own integrity, we will be making every effort within our power to address injustice, prejudice and violence in our own countries and to change the international policies of our own governments. While our own countries are seen as the bullies of the world, it is unsurprising that our message is not always received with favour. Violent conflicts are not just the responsibility of other people, something we can help them with. They are often fuelled, if not caused, by policies, practices and relationships emanating from where we live or consequent on our choices and actions, collectively, past and present. One of the burdens of living in a country where political participation is possible is the moral obligation to use that possibility. The greater our awareness of the need for change, the greater our sense of that obligation becomes, or should become. And where should the first moves be made to create economic justice than in the countries with the greatest economic power? And where should the first moves be made to start dismantling the structures of violence but in those countries that have developed the most powerful means of inflicting it?

Direct Action

There are people in the UK and elsewhere who are committed to acting to fulfil their personal responsibility for dismantling the structures of violence, for instance, those involved in the Trident Ploughshares campaign, who regularly enter military bases and physically 'disarm' weaponry and accompanying technology (Monbiot, 2001). Women in Black protest against violence in different parts of the world, and in Palestine are, as I write, attempting to shield local residents from attack by their presence in their homes. In line with Gandhian principles, these campaigners never lose sight of the humanity of those they confront, and their goal is always dialogue and constructive change, but they are prepared to accept disruption and danger in the meantime.

Those at the 'sharp end' of political action are backed by far more who hold vigils, organise meetings, write letters, collect signatures and lobby political representatives for changes in the policies and actions of their governments, trying to awake their fellow citizens to the need for change.

Not only local but international movements exist to support both direct action and campaigning for an end to militarism and nonviolent approaches to human security, for instance, War Resisters International, the International Fellowship of Reconciliation, and Peace Brigades International. These movements are too easily dismissed as the home of dreamers. They have a serious and urgently needed message and a wealth of thought and experience to contribute, and themselves deserve support.

Influencing Policy: From 'Defence' to Security

Paradoxically, war's consequences are becoming increasingly unacceptable to those who use and threaten it most. Militarism is being eroded from its heart by the growing unwillingness of those countries with the power to coerce their global neighbours to accept loss of life on their own side. The horror of war fits so ill with Western expectations of safety and comfort that soldiers who have been involved in actual fighting and who have seen killing and death close up have been suing for compensation for post-traumatic stress. The military powers are increasingly forced to confine their efforts to bombing campaigns that destroy much but achieve little, causing civilian casualties which are also politically embarrassing; or to the policing of agreements already made – which could, arguably, be done without arms, if enough personnel and resources were made available. In a recent BBC radio interview a British Brigadier assured his interviewer that any NATO troops sent to Macedonia to collect arms from the Albanian rebels would carry only personal arms, and would not be sent at all if there were any chance of a need for armed action. Those intentions may well have been contradicted by events by the time this book is published, but such reluctance to engage in conventional military combat may mean that the time really is ripe for change. The devastating challenge to military-based systems of security which the massive terrorist attacks of September 2001 represented must surely make us re-examine our understanding of human security as such, and see that whatever degree of safety we can hope to achieve will necessarily be based on co-operation rather than threat.

I believe that those of us who wish to contribute to the transformation of violent conflicts and the ways in which they are addressed need to learn from all the experiments and experiences which have been and are being accumulated in nonviolent alternatives at all levels, and to document and publicise them. We need to work to develop new approaches to conflict in our education systems, in the policy and practice of 'law and order', and in popular culture. We need those of us who have some influence in policy-making circles (academics, for instance) to seek out every opportunity to meet with MPs, government members, party leaderships, civil servants and the like. We need to challenge and supplant the idea that national self-interest is the most appropriate basis for foreign policy (perhaps challenge the notion of 'foreign' policy itself). We need to support organisms designed to promote the ideas of mutual support and responsibility among all people, particularly the UN (albeit in need of radical reform) and, within Europe, the OSCE. We need to praise any constructive, ethical actions and policies of governments, building on them as growing points, while at the same time identifying and challenging contradictions.

Ten years of concerted nonviolent action by the Albanian population in Kosovo/a were effectively ignored by the 'international community'. We need to raise public awareness and intensify our own capacities for advocacy and solidarity so that such neglect is never again possible. We need to overcome the fascination of violence and awaken public interest in and support for the alternatives. We need a massive switch from warfare to welfare in the international arena, from military 'security' to common security. The big powers should show their own commitment to overcoming violence and upholding human rights by undertaking a major programme in arms reduction and dismantling their nuclear arsenals.

The new models of community advocated above will come with the recognition of interdependence and a shift from defendedness (dynamically inseparable from aggression), in which others are seen as a threat, to a focus on security, in which others are seen as necessary partners. Although that sentence was easily written, I recognise that its realisation seems light years away. Still we must think the unimaginable, since it is in that way that we create new possibilities. If we consider how rapidly our cultures are changing anyway, we will see how fast and vast cultural change can be.

Similarly, those who oppose all war and wish for a radical departure from militarism are faced with the need for immediate violent crises to be settled, but at the same time know that every time violent means are used to quell violence, the cycle is perpetuated. The nonviolent management of crises is another area calling for urgent research and experimentation, with realistic resources (comparable to those devoted to military 'solutions') devoted to it.

CLOSING THOUGHTS

It will not help to underestimate the difficulties of transforming the culture of violence, or of moving to radically different paradigms of power, or of managing the present in ways that lead to a different kind of future. It is part of the conundrum that each step has to be taken before the necessary conditions are in place. The biggest challenge to the theories of conflict transformation is to think how they can be translated into practice in situations where they have not been applied. 'I wouldn't have started from here' is a reasonable thing to say, but it does not answer the need to advance.

In a postmodern culture, 'emancipatory projects' are unfashionable, redolent of dangerous and unwarranted certainties. Belief in universal values is also frowned upon. While I share in distrust for the kind of modernist certainties and arrogance that have contributed to Western, globalising philosophies and domination, I believe that what is needed by way of response is not the abandonment of all values and aspirations, or the denial of personal and collective power and responsibility in relation to the welfare of other human beings and of our planet. While it is important to be aware of cultural assumptions and sensitive to cultural differences, in a world that seems at the same time smaller and more divided than ever, there is a crying need for the philosophical and practical affirmation of respect as a universal value, without which we undermine our own humanity and any prospect of peaceful coexistence. While respecting separate identities, we need, in the words of Louis Gates Jnr (1994), to move towards a 'politics of identification'.

The day is a long way off where statues in town squares will be not of warriors but of children playing and where flag poles are non-existent or carry the flag of humanity or of the planet. Nonetheless, we can choose to live the future now, in ways within our power, contributing to the development of societies that respect both human weaknesses and strengths, vulnerability and power, commonalities and differences.

The kind of activities described in this book and the actions they support are not a panacea for war-torn societies, but they can be growing points for change, and a source of support for those who, in the words of Adrienne Rich (1984: 264):

age after age, perversely,
with no extraordinary power,
reconstitute the world.

It is to those people that I dedicate this book.

Bibliography

Abu-Nimer, Mohammed. *Conflict Resolution Training in the Middle East: Lessons to be Learned*. (Guildford College, USA: Paper for the Anthropology/ Sociology Department, 1997).

Alexander, Titus. *Unravelling Global Apartheid. An Overview of World Politics*. (Cambridge: Polity Press, 1996).

Anderson, Benedict. *Imagined Communities: Reflections on the Origin and Spread of Nationalism*. (London: Verso Books, 1983).

Anderson, Mary B. *Do No Harm: Supporting Local Capacities for Peace through Aid*. (Cambridge, MA: Local Capacities for Peace Project, 1996).

Assefa, Hizkias. *Peace and Reconciliation as a Paradigm*. (Nairobi: Nairobi Peace Initiative, 1993).

Avruch, Kevin. *Culture and Conflict Resolution*. (Washington, DC: United States Institute of Peace Press, 1998).

Babbit, Eileen F. 'Contributions of Training to International Conflict Resolution'. In Zartman, William I. and Rasmussen, J. Lewis (eds), *Peacemaking in International Conflict: Methods and Techniques*. (Washington, DC: United States Institute of Peace Press, 1997, pp. 365–87).

Berlin, Isaiah. *Concepts and Categories*. (London: Hogarth Press, 1978).

—— 'My Intellectual Path', *New York Review of Books*, 14 May 1998, pp. 53–60.

Billig, M. *Arguing and Thinking*. (Cambridge: Cambridge University Press, 1987).

Bose, Nirmal Kumar. *Selections from Gandhi*. (Ahmedabad: Navajivan Publishing House, 1972).

Boulding, Kenneth. *Ecodynamics*. (London: Sage, 1978).

Burton, J. *Resolving Deep-Rooted Conflicts: A Handbook*. (Lanham, MD: University Press of America, 1987).

—— (ed.) *Conflict: Human Needs Theory*. (London: Macmillan, 1990).

Camplisson, Joe and Hall, Michael. *Hidden Frontiers. Addressing Deep-Rooted Violent Conflict in Northern Ireland and the Republic of Moldova*. (Newtonabbey, County Antrim, Northern Ireland: Island Publications, 1996).

Caritas. *Working for Reconciliation: A Caritas Handbook*. (Vatican City: Caritas Vaticanalis, 1999).

Clark, Howard. *Civil Resistance in Kosovo*. (London: Pluto Press, 2000).

Conciliation Resources. *Accord: An International Review of Peace Initiatives* (Series). (Conciliation Resources, 173 Upper Street, London NI 9WT. www.c-r.org).

Cornelius, Helena and Faire, Shoshona. *Everyone Can Win: How to Resolve Conflict*. (East Roseville, New South Wales: Simon and Schuster, 1989).

Curle, Adam. *Making Peace*. (London: Tavistock Publications, 1971).

—— *In the Middle*. (Leamington Spa: Berg, 1986.) Available on the CCTS website: <www.c-r.org/ccts>.

Duffield, M. *Evaluating Conflict Resolution: Context, Models, and Methodology.* A discussion paper for the Chr. Michelsen Institute. (Bergen, Norway, 1997).

Duryea, Michelle LeBaron. *Conflict and Culture: A Literature Review and Bibliography.* (Canada: UVic Institute for Dispute Resolution, Canada, 1992).

Dworkin, Andrea. *Scapegoat: The Jews, Israel and Women's Liberation.* (London: Virago, 2000).

Eisler, Riane. *The Chalice and the Blade: Our History, Our Future.* (London: Unwin Paperbacks, 1990).

Esman, Milton J. 'Can Foreign Aid Mitigate Conflict?' *Peacework*, no. 13. (Washington, DC: United States Institute of Peace Press, March, 1997).

Fisher, Roger and Ury, William. *Getting to Yes: Negotiating Agreement Without Giving In.* (London: Hutchinson Business Books, 1981).

Fisher, Ronald J. 'The Potential for Peacebuilding: Forging a Bridge from Peacekeeping to Peacemaking'. *Peace and Change*, vol. 18 (1993, pp. 247–66).

—— 'Training as a Form of Interactive Conflict Resolution in Divided Societies'. Paper presented at the Annual Scientific Meeting of the International Society of Political Psychology (Washington, DC, 5–9 July 1995).

—— *Interactive Conflict Resolution.* (New York: Syracuse University Press, 1997).

Fisher, Simon; Ibrahim Abdi, Dekha; Ludun, Jawed; Smith, Richard; Williams, Steve and Williams, Sue. *Working with Conflict* (a manual produced by Responding to Conflict). (London and New York: Zed Books, 2000).

Francis, Diana. 'Respect in Cross-cultural Conflict Resolution Training'. Unpublished doctoral thesis (University of Bath, 1998).

—— (a) 'Conflict Transformation: From Violence to Politics'. (London: Paper written for the Committee for Conflict Transformation Support, June 2000 and printed in CCTS Newsletter No. 9, Summer 2000.) Available on the CCTS website: <www.c-r.org/ccts>.

—— (b) 'Culture, Power Asymmetries and Gender in Conflict Transformation'. In *The Berghof Handbook for Conflict Transformation.* (Berlin: Berghof Research Center for Constructive Conflict Management, 2000.) Available on the Berghof website: <www.berghof-center.org/handbook/>.

—— *Lessons from Kosovo/a: Alternatives to War.* (London: Quaker Peace and Social Witness, 2001).

Francis, Diana and Ropers, Norbert. *Peace Work by Civil Actors in Post-Communist Societies.* (Berghof Occasional Paper No. 10.) (Berlin: Berghof Research Center for Constructive Conflict Management, 1997).

Freire, Paulo. *Pedagogy of the Oppressed.* (London: Penguin, 1972).

Galtung, Johan. 'Cultural Violence'. *Journal of Peace Research*, vol. 27, no. 3 (1990, pp. 291–305).

—— *Peace by Peaceful Means: Peace and Conflict, Development and Civilization.* (London: Sage Publications, 1996).

Gandhi, M. K. *Non-Violence: Weapon of the Brave.* (New Delhi: Orient Paperbacks, no date).

—— *All Men Are Brothers: Autobiographical Reflections.* Compiled and edited by Krishna Kriplani. (New York: Continuum Publishing Corporation, 1980).

Gates Jnr, Henry Louis. 'A Liberalism that Dares to Speak its Name'. *International Herald Tribune*, 30 March 1994.

Gilligan, C. *In a Different Voice*. (Cambridge, MA: Harvard University Press, 1982).

Glasl, Friedrich. *Konfliktmanagement: Ein Handbuch für Führungskräfte, Beraterinnen und Berater*. (Bern: Reies Geistleben, 1997).

Goss-Mayr, Jean and Hildegard. *The Gospel and the Struggle for Peace*. (Alkmaar, The Netherlands: International Fellowship of Reconciliation, 1990).

Gray, John. *Berlin*. (London: Fontana, 1995).

Hall, Michael. *Conflict Resolution: The Missing Element in the Northern Ireland Peace Process*. (Newtown Abbey, County Antrim, Northern Ireland: Island Publications, 1999).

Hart, H. L. A. *The Concept of Law*. (Oxford: Clarendon Press, 1961).

Hoffman, Eva. *Lost in Translation: A Life in a New Language*. (London: Minerva, 1991).

Hopkins, Gerard Manley, in Gardner, W.H. (selection) *Gerard Manley Hopkins*. (Harmondsworth, Middlesex: Penguin Poets, 1953).

International Alert (consultant editor Ian Doucet). *Resource Pack for Conflict Transformation*. (London: International Alert, 1996).

Jelfs, Martin. *Manual for Action*. (London: Action Resources Group, 1982).

Kelman, H. and Cohen, S. 'The Problem-Solving Workshop: a social-psychological contribution to the resolution of international conflicts'. *Journal of Peace Research*, vol. 13 (2) (1976, pp. 79–90).

King, Corretta. *My Life with Martin Luther King*. (Paris: Stock, 1969).

King, Martin Luther. *Why We Can't Wait*. (New York: New American Library, 1963).

Kraybill, Ron. 'Peacebuilders in Zimbabwe'. Unpublished PhD thesis (Eastern Mennonite University, Harrisonburg, USA, 1996).

Kriesberg, Louis. *Constructive Conflicts: From Escalation to Resolution*. (Lanham, MD: Rowman & Little, 1998).

Large, Judith. *The War Next Door*. (Stroud: Hawthorn Press, 1997).

—— 'The Interplay of Domestic, Regional and International Forces in Peacebuilding'. (London: Paper written for the Committee for Conflict Transformation Support, June 2000 and printed in CCTS Newsletter No. 13, Summer 2001.) Available on the CCTS website: <www.c-r.org/ccts>.

LEAP Confronting Conflict and the National Youth Agency (UK). *Playing With Fire: Creative Conflict Resolution for Young Adults*. (New Society Press, 1995).

Lederach, John Paul. *Building Peace – Sustainable Reconciliation in Divided Societies*. (Tokyo: United Nations University Press, 1994).

—— *Preparing for Peace: Conflict Transformation Across Cultures*. (New York: Syracuse University Press, 1995).

Max-Neef, Manfred. 'Reflections on a Paradigm Shift in Economics'. In Inglis, Mary and Kramer, Sandra (eds), *The New Economic Agenda*. (Inverness: Findhorn Press, 1985, p. 147).

Mediation UK. *Training Manual in Conflict Mediation Skills*. (Bristol: Mediation UK, 1995).

Mennonite Conciliation Service. *Mediation and Facilitation Training Manual: Foundations and Skills for Constructive Conflict Transformation* (third edition) (Akron, PA: Mennonite Conciliation Service, 1997).

Miall, Hugh; Ramsbottam, Oliver, and Woodhouse, Tom. *Contemporary Conflict Resolution.* (Cambridge: Polity Press, 1999).

Mitchell, Christopher R. *The Structure of International Conflict.* (Basingstoke and London: Macmillan, 1981).

—— 'Recognising Conflict'. In Woodhouse, Tom (ed.), *Peacemaking in a Troubled World.* (Oxford: Berg, 1991).

—— 'The Process and Stages of Mediation: Two Sudanese Cases'. In Smock, David R. (ed.), *Making War and Waging Peace: Foreign Intervention in Africa.* (Washington, DC: United States Institute of Peace Press, 1993, pp. 142, 147).

Mitchell, Christopher R. and Banks, Michael. *Handbook of Conflict Resolution: The Analytical Problem-Solving Approach.* (New York: Cassell; London: Pinter, 1996).

—— *Handbook of Conflict Resolution; Working for Reconciliation: A Caritas Handbook.* (Vatican City: Caritas, 1999).

Monbiot, George. 'Hell's Grannies'. *Guardian,* 14 August 2001.

Naess, Arne. 'A systematization of Gandhian ethics of conflict resolution'. *Conflict Resolution* (vol. 2, no. 2, 1958).

Narayan, Shriman (ed.). *The Selected Works of Mahatma Gandhi,* vol. 4 (*Constructive Programme*). (Ahmedabad: Navajivan Publishing House, 1968).

Nasrim, Taslima. 'A Disobedient Woman'. *New Internationalist* (no. 289, April 1997).

Nederveen Pieterse, Jan P. *Empire and Emancipation. Power and Liberation on a World Scale. (*New York: Praeger, 1989).

Powers, Roger S. and Vogele, William B. (eds). *Protest, Power and Change: An Encyclopaedia of Nonviolent Action from ACT-UP to Women's Suffrage.* (New York: Farland, 1997).

Quaker Peace Centre, Capetown. *The South African Handbook of Education for Peace.* (Capetown: Quaker Peace Centre, 1992).

Reimann, Cordula. 'Engendering the Field of Conflict Management: Why Gender Does *Not* Matter! Thoughts from a Theoretical Perspective'. In *Peace Studies Papers, Working Paper 2, Fourth Series.* (University of Bradford, Department of Peace Studies, January 2001).

Responding to Conflict. *Working with Conflict.* (London and New York: Zed Books, 2000).

Rich, Adrienne. *The Fact of a Doorframe: Poems Selected and New 1950–1984.* (New York and London: W.W. Norton, 1984).

Rockefeller, Steven C. *John Dewey: Religious Faith and Democratic Humanism.* (New York: Columbia University Press, 1994).

Ropers, Norbert. *Peaceful Intervention: Structures, Processes and Strategies for the Constructive Regulation of Ethno-Political Conflicts.* (Berghof Report No. 1.) (Berlin: Berghof Research Center for Constructive Conflict Management, 1995).

Rothman, J. *From Confrontation to Co-operation: Resolving Ethnic and Regional Conflict.* (Newbury Park, CA: Sage Publications, 1992).

Rouhana, Nadim N. 'Unofficial Third-Party Intervention in International Conflict: Between Legitimacy and Disarray'. *Negotiation Journal* (July 1995, pp. 258, 262, 266).

Sahgal, Gita and Yuval-Davis, Nira (eds). *Refusing Holy Orders: Women and Fundamentalism in Britain*. (London: Virago Press, 1992).

Saint-Exupery, Antoine de. *Wind, Sand and Stars. (*First published in French in 1939 as *Terre des Hommes*.) (London: Penguin, 1995).

Salem, Paul. 'A Critique of Conflict Resolution from a Non-Western Perspective'. In Salem, Paul (ed.), *Conflict Resolution in the Arab World: Selected Essays*. (Beirut: American University of Beirut, Lebanon, 1997, pp. 11–24).

Scott, James. *Domination and the Arts of Resistance: Hidden Transcripts*. (Yale and London: Yale University Press, 1990).

Sharp, Gene. *The Politics of Nonviolent Action*. (Boston: Porter Sargent, 1973).

Taylor, Charles. 'The Politics of Recognition'. In Gutmann, Amy (ed.), *Multi-culturalism*. (Princeton, NJ: Princeton University Press, 1994, pp. 25–74).

Williams, Sue and Steve. *Being in the Middle by Being at the Edge: Quaker Experience of Non-official Political Mediation*. (London: Quaker Peace and Service, 1994).

Woodhouse, Tom (ed.). *Peacemaking in a Troubled World*. (Oxford: Berg, 1991).

Index